LAKE SUPERIOR

Rocks & Minerals

A Field Guide to the Lake Superior Area

Bob Lynch & Dan Lynch

Adventure Publications, Inc.
Cambridge, MN

Dedication

To Nancy Lynch, wife and mother, for her help in collecting data and her support for our book

Acknowledgments

Thanks to the following for providing specimens and information: George Robinson, Ph.D., 3M Museum, Lake County Historical Society, Karen Brzys, Ken Flood, Keith and Teresa Bartel, David Gredzens, Karl Wanink, John Heikkinen, Robert Weikert, and Tom Bjugstad.

Special thanks to the A.E. Seaman Mineral Museum at Michigan Technological University in Houghton, Michigan, for furnishing specimens and information.

Initial editing by Bob Lynch Jr.

Photography by Dan Lynch

Cover and book design by Jonathan Norberg

Edited by Brett Ortler

10 9 8 7 6 5 4 3
Copyright 2008
Published by Adventure Publications, Inc.
820 Cleveland St. S
Cambridge, MN 55008
1-800-678-7006
www.adventurepublications.net
Printed in China
ISBN-13: 978-1-59193-095-2
ISBN-10: 1-59193-095-2

Table of Contents

The Lake Superior region

Lake Superior

Minnesota

Wisconsin

Michigan

Introduction

Few places in the world are as geologically diverse as the Lake Superior Basin. Millions of years of volcanic and glacial activity have cracked, crushed, pushed, and pulled this area into a veritable gold mine of collectible rocks and minerals. As a result, the immense copper deposits in Michigan, the iron-rich areas in Wisconsin, and the prime agate collecting in Minnesota all make Lake Superior a top destination for rock hounds.

After years of researching the rocks and minerals found in the three Lake Superior states, we've written this book to help our fellow rock collectors and beachcombers figure out what they're finding. Whether you're searching the lakeshore, riverbeds, gravel pits, or the overburden piles found near old mines, you can read about the most commonly found materials in this book. It is important to know that while many of these rocks and minerals are found throughout the world, some of the tips and tricks we provide to finding these specimens are not universal; they are what you need to know to find them in the Lake Superior area.

Rocks vs. Minerals

Many people hunt for rocks and minerals without knowing the difference between the two. It's simple: A mineral is a crystallization or solidification of a compound, whereas a rock is a combination of minerals. Because rocks can contain different amounts of the same minerals from specimen to specimen, hardness and streak tests do not apply to rocks.

Protected Places

All too often people are caught hunting for rocks and minerals in places they shouldn't be and get into trouble. When going out to hunt for specimens, you need to be absolutely sure the places you are looking are not on private property. Much of Lake Superior's shorelines are privately owned, and rock hunting there is illegal unless you have explicit permission from the landowner. The government protects some areas, like Isle Royale National Park and Michigan's Picture Rocks National Lakeshore, and it is illegal to remove anything, including rocks, from such sites. Other places that are often, if not always, privately owned are gravel pits and mine dumps. It is also important to note that many gravel pits and mine dumps, no matter how unused they may be, can be very dangerous places. The best way to prepare for a rock hunting excursion is to simply be aware.

Cleaning Minerals

One of the most common questions about rocks and minerals is how to clean them—especially agates. Many people think putting rocks and minerals in acid will clean them better than anything else. This practice is very dangerous, and other methods work just as well. Soap and warm water are the best things to try. With agates, oven cleaner will work well to brighten the color of the rock. If you want to give a specimen a wet look, a very light coating of mineral oil will do the trick.

Hardness and Streak

There are two indispensable techniques everyone wishing to identify minerals should know about: hardness and streak tests. Nearly all minerals exhibit these two traits, which can help you determine what they are.

Hardness is the measure of how resistant a mineral is to abrasion. The hardness scale ranges from 1 to 10, with 10 being the hardest. A mineral with a hardness of 1 is talc, a very chalk-like stone that can easily be scratched by your bare hands. An example of a mineral with a hardness of 10 is diamond, which is the hardest natural substance on Earth. Minerals of the Lake Superior region all fall somewhere in the middle, so learning how to perform a scratch test is key. Using your fingernail, a coin, a piece of glass, or a knife to scratch a stone are common methods to determine hardness. There are also hardness kits you can purchase that have a tool of each hardness. On page 8, you'll find a chart that shows which tools will scratch a mineral of a particular hardness.

The second test you'll need to familiarize yourself with is streak. Most minerals have a distinct color when crushed and turned to powder. A mineral's powder color is the same as its streak color. Often the powder is a completely different color than the stone. Hematite, for example, is dark metallic gray or black in color, but its streak is rusty red in color. A mineral's streak is easily determined by rubbing it along a special streak plate or a piece of hard, unfinished porcelain. The only rule to keep in mind for this test is that a mineral harder than the streak plate won't have a streak. Instead, it will actually scratch the plate itself.

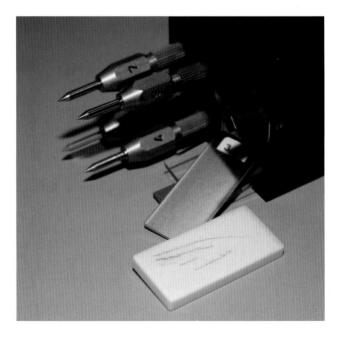

The Mohs Hardness Scale

The Mohs Hardness Scale is the primary measure of mineral hardness. This scale ranges from 1 to 10, from softest to hardest. Ten minerals commonly associated with the scale are listed here. Also listed are common tools and their hardness for use in doing a scratch test on a mineral. If a mineral is scratched by a tool, you know it is softer than that particular hardness.

HARDNESS	EXAMPLE MINERAL	TOOL
1	Talc	
2	Gypsum	
2.5		Fingernail
3	Calcite	
3.5		Copper Coin
4	Fluorite	
5	Apatite	
5.5		Glass, Steel Nail
6	Orthoclase	
6.5		Streak Plate
7	Quartz	
8	Topaz	
9	Corundum	
10	Diamond	

For example, if a mineral is scratched by a copper coin but not your fingernail, you can conclude that its hardness is 3, equal to that of calcite. If a mineral is harder than 6.5, or the hardness of a streak plate, it will have no streak unless weathered to a softer state.

Quick Identification Guide

Use this quick identification guide to help you determine which rock you have found. We've listed the primary color groups and characteristic traits of rocks and minerals of the Lake Superior region, as well as the page number where you can read more about your possible find. The colors and traits listed here are simply the most common for each stone. Not every rock you find will conform to these characteristics.

If white and...	then try...
White, ball-shaped crystals growing on or in other rock	analcime, page 155
Soft, easily scratched translucent crystals, nodules, or amygdules	calcite, page 53
Soft, chalky, white stone	limestone, page 173
Very hard, translucent crystals, nodules, or amygdules	quartz, page 97
Dense, grainy stones with the texture of sand	quartzite, page 103
Brittle, soft, easily scratched fibrous crystals in nodules	zeolite, page 113

WHITE

Quick Identification Guide (continued)

GRAY

If gray and...	then try...
Gray green rock with some translucent areas	anorthosite, page 123
Extremely common, dense gray stone often found on beaches	basalt, page 47
Light gray cauliflower-shaped nodules with colored interior	datolite, page 159
Gray rock containing lighter, flower-shaped spots	diabase, page 67
Dark rock containing reflective or translucent areas	gabbro, page 71
Layered gray rock found in flat, sheet-like pieces	shale, page 109

BLUE

If blue and...	then try...
Blue green masses within other rock	chrysocolla, page 157

13

Quick Identification Guide (continued)

BLACK

If black and...	then try...
Heavy, black mineral with a metallic sheen and yellow oxidation	goethite, page 73
Black mineral with a metallic sheen and red oxidation	hematite, page 77
Black, metallic mineral that is magnetic	magnetite, page 85
Dark, layered, reflective crystals growing on or in other rock	mica, page 87
Black, metallic, fan-shaped crystals	pyrolusite, page 95 or manganite, page 175

YELLOW

If yellow and...	then try...
Very hard rock with a waxy appearance	chert, page 57

Quick Identification Guide (continued)

BROWN

	If brown and...	then try...
	Light brown to red rounded stones with the appearance of being sculpted	concretions, page 145
	Short yellow brown fibrous crystals within another rock	grunerite, page 165
	Pink brown rock with a cracked appearance	Kona dolomite, page 171
	Grainy, red brown to yellow brown coarse-textured rock with the appearance of sand	sandstone, page 107
	Extremely fine-grained, yellow brown to dark brown rock	siltstone, page 111
	Small, hard crystals, often in a cross shape	staurolite, page 187

ORANGE

	If orange and...	then try...
	Hard, bright orange crystals or nodules within other rocks	feldspar, page 69
	Soft, light orange to pink nodules easily scratched by your fingernail	laumontite, page 83

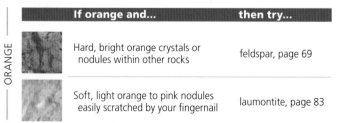

Quick Identification Guide *(continued)*

GREEN

	If green and...	then try...
	Dark green nodules easily scratched by a coin	chlorite, page 59
	Deep green, very dense rock with a mottled appearance	Ely greenstone, page 127
	Dark, brittle, yellow green crystals	epidote, page 161
	Pale green, small nodules with a frosted appearance	lintonite, page 133
	Light, pale green, semi-translucent hard crystals	prehnite, page 91
	Dark green chatoyant nodules with a "turtle-back" pattern	pumpellyite, page 181
	Yellow green, striated, waxy crystals	serpentine, page 183

Quick Identification Guide (continued)

VIOLET OR PINK

If violet or pink and...	then try...
Translucent purple crystal points or masses	amethyst, page 121
Pink or white fibrous crystal nodules in basalt	thomsonite, page 139
Pink or white fibrous crystals or nodules	zeolite, page 113

RED

If red and...	then try...
Translucent reddish brown nodules with a waxy texture	chalcedony, page 55
Deep red or purple red opaque stones	jasper, page 79
Red fine-grained rock often with many vesicles	rhyolite, page 105

Quick Identification Guide (continued)

METALLIC

If metallic and...	then try...
Golden crystals, often with a rainbow-colored sheen	chalcopyrite, page 143
Reddish metal with greenish blue oxidation	copper, page 63
Copper and silver contained together in a single specimen	halfbreed, page 167
Gold crystals with a greenish hue and bladed construction	marcasite, page 177
Silver-bronze metal, often with "flowers" of quartz within	mohawkite, page 179
Golden cubic crystals	pyrite, page 93
Gray metallic mineral with black oxidization	silver, page 185

Quick Identification Guide (continued)

MULTICOLORED OR BANDED

	If multicolored or banded and...	then try...
	Translucent banded stone with reds, browns, and whites	agate, page 19
	Fibrous stone containing black, yellow, red, white and gold	binghamite, page 125 or silkstone, page 135
	Rock made up of many smaller rounded stones	conglomerate, page 61
	White, gray, pink or red rock with flecks of black or other colors	granite, page 75
	Layered, mushroom-shaped structures within dense red or green material	mary ellen jasper, page 131
	Complete crystals embedded within a different dense rock	porphyry, page 89

Sample Page

HARDNESS: 7 **STREAK:** White

AREA: The states where this rock or mineral can be found

Lake Superior Area
Occurrence

ENVIRONMENT: The type of place where this rock or mineral can be found. For the purposes of this book, the primary environments listed include the lakeshore, riverbeds, gravel pits, and mine dumps, referring to the overburden piles left at the sites of many mines.

WHAT TO LOOK FOR: Common and characteristic traits of the rock or mineral, the first sample photo of which will be shown on the preceding page

SIZE: The general size range of the rock or mineral

COLOR: The general colors a rock or mineral exhibits in its natural state

OCCURRENCE: How easy or difficult this rock or mineral is to find. "Very common" means the material takes virtually no effort to find. "Common" means the material can be found with little effort. "Uncommon" means the material may take a good deal of hunting to find. "Rare" means the material will take great lengths of time and energy to find. "Very rare" means the material is so uncommon that you will be lucky to even find a trace of it.

NOTES: These are additional notes about the rock or mineral that include how to find it, where to find it, and interesting facts about it.

WHERE TO LOOK: Here you'll find specific places and geographical locations where you can find the rock or mineral.

Agate in basalt

Typical agates found on a beach

Agate

HARDNESS: 7 **STREAK:** White

Lake Superior Area
Occurrence

Agate is the most collectible and valuable of all of Lake Superior's native rocks and minerals. These stones are a banded form of chalcedony, and can be a wide range of colors. The most common are red, brown, gray, white, and rarely green, blue, and yellow. The red variety is the most sought after, and it gets its color from an iron oxide which stains it red.

Lake Superior agates were worn out of their host rock by the many glaciers that moved through this area during the ice ages. These glaciers pulverized the surrounding basalt and freed the agate, which resisted weathering due to its hardness. However, the agates did not survive the glaciers unscathed. Most agates exhibit banding. Since the agates formed as rock nodules, with the bands contained inside, agates found with visible banding have their interiors exposed. These were broken at some point, most likely by the glaciers.

The same glaciers that carved the agates out of their host rock transported them all over the region. Lake Superior agates can be found all over Minnesota, Wisconsin, and Upper Michigan. Some agates have been found even as far south as Iowa. As the agates were pushed everywhere by the glaciers, they can be found in gravel pits, riverbeds, and on the lakeshore. There have also been instances where iron or copper mines have unearthed agates.

Agates can be extremely valuable. Color, size, and overall quality are determining factors for assessing value. The most sought after colors by serious collectors are high-contrast, red and white banded agates. Agates larger than an adult's fist are also widely desired, and the more solid the agate, the better. In this book we'll discuss the many varieties of agate separately.

Broken pieces of banded agate

Quartz

Agate, Brecciated

HARDNESS: 7 **STREAK:** White

**Lake Superior Area
Occurrence**

AREA: Minnesota, Wisconsin, and Michigan

ENVIRONMENT: Lakeshore, gravel pits, mine dumps, and riverbanks

WHAT TO LOOK FOR: Translucent, banded stones of varying colors, often with quartz crystals

SIZE: Lake Superior agates can be any size, from tiny grains of sand up to softball-sized specimens.

COLOR: Agates can run the full range of colors, but the most common colors are red, brown, gray, or white with clear, translucent quartz.

OCCURRENCE: Rare

NOTES: Brecciated agate forms the same way other types of breccia (broken, then recombined rock) form. A normally formed agate is crushed and broken and its pieces were then cemented back together by silica and quartz. The pieces are arranged randomly, but you can often see some of the agate's original banding in the individual pieces. These "destroyed" agates are very rare and can be very valuable.

WHERE TO LOOK: Agates are found along the lakeshore in all three states. Also, they can be found in gravel pits, riverbeds, farm fields, and along highways and dirt roads. Some collecting areas: In Minnesota, the beaches from Duluth to Two Harbors, Flood Bay and Burlington Bay in Two Harbors, and Beaver Bay in Beaver Bay, MN. In Michigan, the best bet is the western side of the Keweenaw Peninsula, and from Grand Marais, Michigan east along Lake Superior.

Agate, Eye

HARDNESS: 7 **STREAK:** White

Lake Superior Area
Occurrence

AREA: Minnesota, Wisconsin, and Michigan

ENVIRONMENT: Lakeshore, gravel pits, mine dumps, and riverbanks

WHAT TO LOOK FOR: Translucent, banded stones of varying colors, often with quartz crystals

SIZE: Lake Superior agates can be any size, from tiny grains of sand up to softball-sized specimens.

COLOR: Agates can run the full range of colors, but the most common colors are red, brown, gray, or white with clear, translucent quartz.

OCCURRENCE: Uncommon

NOTES: If you look carefully at many agates, especially those with very smooth, complete nodular (round) structures, you may notice perfectly round patterns on or beneath the surface. These have become known as agate "eyes" and are more common than most collectors think. The most collectible and valuable agates have eyes with concentric banding which resemble a target pattern. The most sought after of all have eyes that are dime-sized or larger, though agate eyes of this size are very rare.

WHERE TO LOOK: Agates are found along the lakeshore in all three states. Also, they can be found in gravel pits, riverbeds, farm fields, and along highways and dirt roads. Some collecting areas: In Minnesota, the beaches from Duluth to Two Harbors, Flood Bay and Burlington Bay in Two Harbors, and Beaver Bay in Beaver Bay, MN. In Michigan, the best bet is the western side of the Keweenaw Peninsula, and from Grand Marais, Michigan east along Lake Superior.

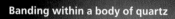
Banding within a body of quartz

Agate, Floater

HARDNESS: 7 **STREAK:** White

Lake Superior Area
Occurrence

AREA: Minnesota, Wisconsin, and Michigan

ENVIRONMENT: Lakeshore, gravel pits, mine dumps, and riverbanks

WHAT TO LOOK FOR: Translucent, banded stones of varying colors, often with quartz crystals

SIZE: Lake Superior agates can be any size, from tiny grains of sand up to softball-sized specimens.

COLOR: Agates can run the full range of colors, but the most common colors are red, brown, gray, or white with clear, translucent quartz.

OCCURRENCE: Common

NOTES: This type of agate is named because its bands appear to be floating in a body of quartz. Often this quartz can itself appear banded, signifying different layers of quartz growth. Sometimes, an agate that at first appears to be what's known as a "quartz ball," or a mass of quartz with an agate shell, can be cut to reveal a beautiful center made up of previously hidden agate banding.

WHERE TO LOOK: Agates are found along the lakeshore in all three states. Also, they can be found in gravel pits, riverbeds, farm fields, and along highways and dirt roads. Some collecting areas: In Minnesota, the beaches from Duluth to Two Harbors, Flood Bay and Burlington Bay in Two Harbors, and Beaver Bay in Beaver Bay, MN. In Michigan, the best bet is the western side of the Keweenaw Peninsula, and from Grand Marais, Michigan east along Lake Superior.

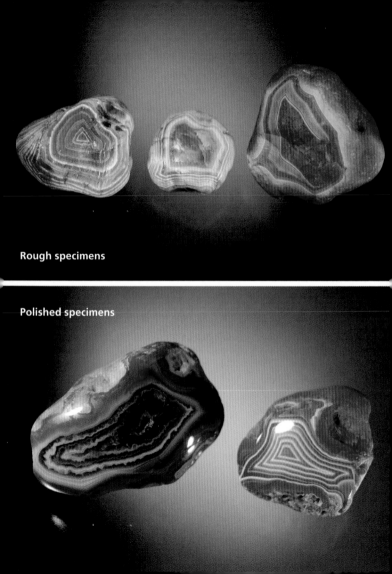

Rough specimens

Polished specimens

Agate, Fortification

HARDNESS: 7 **STREAK:** White

Lake Superior Area
Occurrence

AREA: Minnesota, Wisconsin, and Michigan

ENVIRONMENT: Lakeshore, gravel pits, mine dumps, and riverbanks

WHAT TO LOOK FOR: Translucent, banded stones of varying colors, often with quartz crystals

SIZE: Lake Superior agates can be any size, from tiny grains of sand up to softball-sized specimens.

COLOR: Agates can run the full range of colors, but the most common colors are red, brown, gray, or white with clear, translucent quartz.

OCCURRENCE: Common

NOTES: This variety of Lake Superior agate is one of the most common and most sought after. Fortification agates get their name from their exquisite banding, which begins at the edges of the agate and advances inward and looks like the walls of a fort viewed from above. Alternating red and white banding is by far the most valuable type of agate to collectors, and the best specimens of this coloration can be found in fortification agates. Many fortification agates have a body of quartz in the center of the bands.

WHERE TO LOOK: Agates are found along the lakeshore in all three states. Also, they can be found in gravel pits, riverbeds, farm fields, and along highways and dirt roads. Some collecting areas: In Minnesota, the beaches from Duluth to Two Harbors, Flood Bay and Burlington Bay in Two Harbors, and Beaver Bay in Beaver Bay, MN. In Michigan, the best bet is the western side of the Keweenaw Peninsula, and from Grand Marais, Michigan east along Lake Superior.

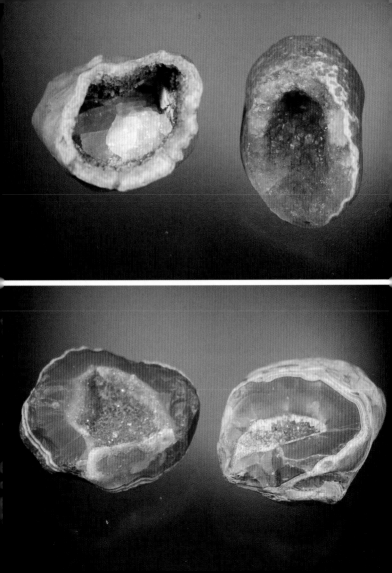

Agate, Geode

HARDNESS: 7 **STREAK:** White

Lake Superior Area
Occurrence

AREA: Minnesota, Wisconsin, and Michigan

ENVIRONMENT: Lakeshore, gravel pits, mine dumps, and riverbanks

WHAT TO LOOK FOR: Translucent, banded stones of varying colors, often with quartz crystals

SIZE: Lake Superior agates can be any size, from tiny grains of sand up to softball-sized specimens.

COLOR: Agates can run the full range of colors, but the most common colors are red, brown, gray, or white with clear, translucent quartz.

OCCURRENCE: Uncommon

NOTES: Agate geodes are far more common in other parts of the world, such as Brazil, but they can still be found in the Lake Superior area. These are essentially agates that weren't "finished" yet. As these agates grew inward, the silica necessary to keep forming the microcrystalline agate material ran out and the remaining spaces were left hollow, often lined with large, coarse quartz crystals.

WHERE TO LOOK: Agates are found along the lakeshore in all three states. Also, they can be found in gravel pits, riverbeds, farm fields, and along highways and dirt roads. Some collecting areas: In Minnesota, the beaches from Duluth to Two Harbors, Flood Bay and Burlington Bay in Two Harbors, and Beaver Bay in Beaver Bay, MN. In Michigan, the best bet is the western side of the Keweenaw Peninsula, and from Grand Marais, Michigan east along Lake Superior.

Rough specimens

Polished specimens

Agate, Moss

HARDNESS: 7 **STREAK:** White

**Lake Superior Area
Occurrence**

AREA: Minnesota, Wisconsin, and Michigan

ENVIRONMENT: Lakeshore, gravel pits, mine dumps, and riverbanks

WHAT TO LOOK FOR: Translucent, banded stones of varying colors, often with quartz crystals

SIZE: Lake Superior agates can be any size, from tiny grains of sand up to softball-sized specimens.

COLOR: Agates can run the full range of colors, but the most common colors are red, brown, gray, or white with clear, translucent quartz.

OCCURRENCE: Uncommon

NOTES: This variety of agate formed very differently than the other, more collectible types. Moss agates formed with many inclusions (minerals like iron oxides, manganese, or hematite that formed within the agate). These other minerals caused the banding to become irregular and resemble moss, giving these agates their name. Moss agates are not nearly as sought after as other agate types, mostly because of their outer appearance, but once cut and polished, moss agates can be just as beautiful as other forms of agate.

WHERE TO LOOK: Agates are found along the lakeshore in all three states. Also, they can be found in gravel pits, riverbeds, farm fields, and along highways and dirt roads. Some collecting areas: In Minnesota, the beaches from Duluth to Two Harbors, Flood Bay and Burlington Bay in Two Harbors, and Beaver Bay in Beaver Bay, MN. In Michigan, the best bet is the western side of the Keweenaw Peninsula, and from Grand Marais, Michigan east along Lake Superior.

Rough specimens

Polished specimens

Agate, Paint

HARDNESS: 7 **STREAK:** White

Lake Superior Area
Occurrence

AREA: Minnesota, Wisconsin, and Michigan

ENVIRONMENT: Lakeshore, gravel pits, mine dumps, and riverbanks

WHAT TO LOOK FOR: Translucent, banded stones of varying colors, often with quartz crystals

SIZE: Lake Superior agates can be any size, from tiny grains of sand up to softball-sized specimens.

COLOR: Agates can run the full range of colors, but the most common colors are red, brown, gray, or white with clear, translucent quartz.

OCCURRENCE: Uncommon

NOTES: Paint agates can be very valuable due to their unique and beautiful colors. What makes them different from other agates is that the concentration of minerals within the bands is much higher than in other agates, making the banding appear very opaque. Minerals in the bands of these agates oxidized (reacted with oxygen) to a greater degree than in other agates. Characteristic colors for paint agates include deep reds, pinks, and oranges, but occasionally blue, gray, green, or yellow appear due to the presence of other minerals.

WHERE TO LOOK: Agates are found along the lakeshore in all three states. Also, they can be found in gravel pits, riverbeds, farm fields, and along highways and dirt roads. Some collecting areas: In Minnesota, the beaches from Duluth to Two Harbors, Flood Bay and Burlington Bay in Two Harbors, and Beaver Bay in Beaver Bay, MN. In Michigan, the best bet is the western side of the Keweenaw Peninsula, and from Grand Marais, Michigan east along Lake Superior.

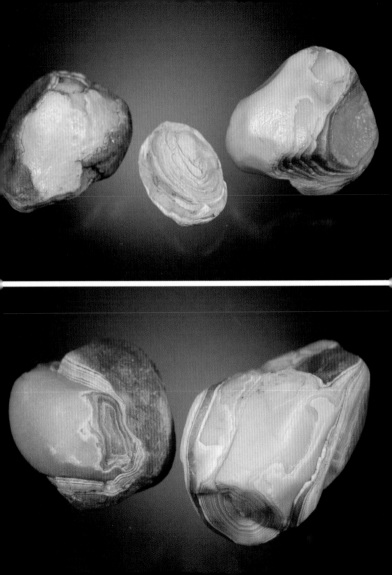

Agate, Peeled

HARDNESS: 7 **STREAK:** White

Lake Superior Area
Occurrence

AREA: Minnesota, Wisconsin, and Michigan

ENVIRONMENT: Lakeshore, gravel pits, mine dumps, and riverbanks

WHAT TO LOOK FOR: Translucent, banded stones of varying colors, often with quartz crystals

SIZE: Lake Superior agates can be any size, from tiny grains of sand up to softball-sized specimens.

COLOR: Agates can run the full range of colors, but the most common colors are red, brown, gray, or white with clear, translucent quartz.

OCCURRENCE: Uncommon

NOTES: Peeled agates can actually be any variety of agate but are different in one way. Peeled agates have actually shed some of their banding. This separation occurred between the bands themselves and gives part of the agate's surface a very smooth, waxy look. This happened when moisture within the agate froze and separated the bands, weakening the bonds with the rest of the stone. More freezing or other weathering finally caused these loose portions to fall off completely.

WHERE TO LOOK: Agates are found along the lakeshore in all three states. Also, they can be found in gravel pits, riverbeds, farm fields, and along highways and dirt roads. Some collecting areas: In Minnesota, the beaches from Duluth to Two Harbors, Flood Bay and Burlington Bay in Two Harbors, and Beaver Bay in Beaver Bay, MN. In Michigan, the best bet is the western side of the Keweenaw Peninsula, and from Grand Marais, Michigan east along Lake Superior.

Crushed material

Intact banding

Specimen courtesy of Keith and Theresa Bartel

Agate, Ruin

HARDNESS: 7 **STREAK:** White

Lake Superior Area
Occurrence

AREA: Minnesota, Wisconsin, and Michigan

ENVIRONMENT: Lakeshore, gravel pits, mine dumps, and riverbanks

WHAT TO LOOK FOR: Translucent, banded stones of varying colors, often with quartz crystals

SIZE: Lake Superior agates can be any size, from tiny grains of sand up to softball-sized specimens.

COLOR: Agates can run the full range of colors, but the most common colors are red, brown, gray, or white with clear, translucent quartz.

OCCURRENCE: Uncommon

NOTES: Ruin agates are much like brecciated agates (agates which are broken, then recombined) but they are much less crushed. These agates underwent great pressure and were crushed but not completely destroyed. Instead, the pressure pulverized a section of the agate and left the other parts intact. Silica cemented the agate back together, and the agates often maintain their original nodular shapes.

WHERE TO LOOK: Agates are found along the lakeshore in all three states. Also, they can be found in gravel pits, riverbeds, farm fields, and along highways and dirt roads. Some collecting areas: In Minnesota, the beaches from Duluth to Two Harbors, Flood Bay and Burlington Bay in Two Harbors, and Beaver Bay in Beaver Bay, MN. In Michigan, the best bet is the western side of the Keweenaw Peninsula, and from Grand Marais, Michigan east along Lake Superior.

Agate, Sagenite

HARDNESS: 7 **STREAK:** White

AREA: Minnesota, Wisconsin, and Michigan

Lake Superior Area
Occurrence

ENVIRONMENT: Lakeshore, gravel pits, mine dumps, and riverbanks

WHAT TO LOOK FOR: Translucent, banded stones of varying colors, often with quartz crystals

SIZE: Lake Superior agates can be any size, from tiny grains of sand up to softball-sized specimens.

COLOR: Agates can run the full range of colors, but the most common colors are red, brown, gray, or white with clear, translucent quartz.

OCCURRENCE: Uncommon

NOTES: Sagenite agates are a unique and uncommon type of agate. When hunting for sagenites, look for radiating (spread out from a central point), fan-shaped arrangements of tube-like structures. These formations occur when agate material replaces long, slender crystals already present in the cavity (rutile, tourmaline, or goethite). This results in needle-like shapes which primarily occur in red, gold, or white, and in different colors than the rest of the agate.

WHERE TO LOOK: Agates are found along the lakeshore in all three states. Also, they can be found in gravel pits, riverbeds, farm fields, and along highways and dirt roads. Some collecting areas: In Minnesota, the beaches from Duluth to Two Harbors, Flood Bay and Burlington Bay in Two Harbors, and Beaver Bay in Beaver Bay, MN. In Michigan, the best bet is the western side of the Keweenaw Peninsula, and from Grand Marais, Michigan east along Lake Superior.

Polished specimen

Rough specimens

Agate, Seam

HARDNESS: 7 **STREAK:** White

Lake Superior Area
Occurrence

AREA: Minnesota, Wisconsin, and Michigan

ENVIRONMENT: Lakeshore, gravel pits, mine dumps, and riverbanks

WHAT TO LOOK FOR: Translucent, banded stones of varying colors, often with quartz crystals

SIZE: Lake Superior agates can be any size, from tiny grains of sand up to softball-sized specimens.

COLOR: Agates can run the full range of colors, but the most common colors are red, brown, gray, or white with clear, translucent quartz.

OCCURRENCE: Uncommon

NOTES: Seam agates are less common than the other types of agates and yet are not widely sought after by collectors. These peculiar agates formed within a crack or break in their host rocks, rather than in vesicles (cavities formed by gas bubbles). Their bands are often straighter or longer, and the agates sometimes contain cavities, cracks, or other unique structures, such as tubes or crystals. Most of these originate because the agate formed from top and bottom of the cavity and filled into the center.

WHERE TO LOOK: Agates are found along the lakeshore in all three states. Also, they can be found in gravel pits, riverbeds, farm fields, and along highways and dirt roads. Some collecting areas: In Minnesota, the beaches from Duluth to Two Harbors, Flood Bay and Burlington Bay in Two Harbors, and Beaver Bay in Beaver Bay, MN. In Michigan, the best bet is the western side of the Keweenaw Peninsula, and from Grand Marais, Michigan east along Lake Superior.

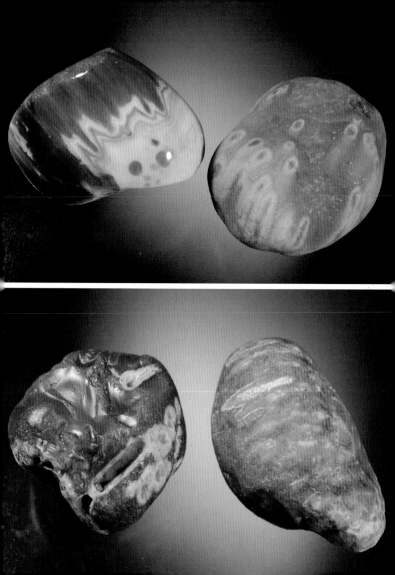

Agate, Tube

HARDNESS: 7 **STREAK:** White

AREA: Minnesota, Wisconsin, and Michigan

Lake Superior Area
Occurrence

ENVIRONMENT: Lakeshore, gravel pits, mine dumps, and riverbanks

WHAT TO LOOK FOR: Translucent, banded stones of varying colors, often with quartz crystals

SIZE: Lake Superior agates can be any size, from tiny grains of sand up to softball-sized specimens.

COLOR: Agates can run the full range of colors, but the most common colors are red, brown, gray, or white with clear, translucent quartz.

OCCURRENCE: Rare

NOTES: Tube structures occur within agates when minerals (such as goethite) form before or during the creation of the agate and the agate develops around these minerals. The agate often coats these stalactitic (icicle-like) mineral structures with multiple bands before the rest of the agate fills in the nodule (round rock cluster). The result is a tube agate. These rare agates can greatly resemble eye agates when the tubes have only been partially exposed and when the rest of the tubes are not visible.

WHERE TO LOOK: Agates are found along the lakeshore in all three states. Also, they can be found in gravel pits, riverbeds, farm fields, and along highways and dirt roads. Some collecting areas: In Minnesota, the beaches from Duluth to Two Harbors, Flood Bay and Burlington Bay in Two Harbors, and Beaver Bay in Beaver Bay, MN. In Michigan, the best bet is the western side of the Keweenaw Peninsula, and from Grand Marais, Michigan east along Lake Superior.

Rough specimens

Polished specimens

Agate, Water Level

HARDNESS: 7 **STREAK:** White

Lake Superior Area
Occurrence

AREA: Minnesota, Wisconsin, and Michigan

ENVIRONMENT: Lakeshore, gravel pits, mine dumps, and riverbanks

WHAT TO LOOK FOR: Translucent, banded stones of varying colors, often with quartz crystals

SIZE: Lake Superior agates can be any size, from tiny grains of sand up to softball-sized specimens.

COLOR: Agates can run the full range of colors, but the most common colors are red, brown, gray, or white with clear, translucent quartz.

OCCURRENCE: Uncommon

NOTES: Water level agates are another unique type of agate that can be found in the Lake Superior area. These agates exhibit parallel bands, sometimes in addition to other agate structures, namely fortifications. These agates form when silica flows into a cavity intermittently rather than as a constant flow. Because of this, the silica settles to the bottom and solidifies, rather than forming bands along the inside of the cavity.

WHERE TO LOOK: Agates are found along the lakeshore in all three states. Also, they can be found in gravel pits, riverbeds, farm fields, and along highways and dirt roads. Some collecting areas: In Minnesota, the beaches from Duluth to Two Harbors, Flood Bay and Burlington Bay in Two Harbors, and Beaver Bay in Beaver Bay, MN. In Michigan, the best bet is the western side of the Keweenaw Peninsula, and from Grand Marais, Michigan east along Lake Superior.

Beach-worn specimens

Rough specimens

Basalt

HARDNESS: 5-6 **STREAK:** N/A

Lake Superior Area
Occurrence

AREA: Extremely abundant in all three states

ENVIRONMENT: Lakeshore, gravel pits, mine dumps, and riverbanks

WHAT TO LOOK FOR: Most commonly black, gray, or bluish dense rock with no visible grains

SIZE: Basalt is an extremely common volcanic rock, which means that basalt can be found as huge masses that extend for miles or as tiny pebbles.

COLOR: Basalt can have a great range of color, from dark gray and black to greens, blues, and reds, but is most commonly gray.

OCCURRENCE: Very common

NOTES: Basalt in this region is mainly a result of the North American continent trying to split in half many millions of years ago. Today, basalt makes up much of Lake Superior's shoreline, including the massive sheets of ledge rock that extend deep into the lake.

Basalt has a vast array of colors that can make it hard to distinguish it from other volcanic rock, such as rhyolite. Basalt, unlike other minerals, has no grain-like structures. As lava, basalt cooled so quickly that it became very dense and so finely grained that even under a microscope you cannot find any distinctive features. This makes basalt unique, as even rhyolite has a distinctive grainy appearance under a microscope.

WHERE TO LOOK: Basalt is very common along Lake Superior; look anywhere there is exposed rock.

Basalt, Amygdaloidal

HARDNESS: 5-6 **STREAK:** N/A

Lake Superior Area
Occurrence

AREA: Extremely abundant in all three states

ENVIRONMENT: Lakeshore, gravel pits, mine dumps, and riverbanks

WHAT TO LOOK FOR: Most commonly black, gray, or bluish dense rock with no visible grains

SIZE: Basalt is an extremely common volcanic rock, which means that basalt can be found as huge masses that extend for miles or as tiny pebbles.

COLOR: Basalt can have a great range of color, from dark gray and black to greens, blues, and reds, but is most commonly gray.

OCCURRENCE: Very common

NOTES: Amygdaloidal basalt is vesicular basalt that has received many secondary minerals that flowed into its vesicles (cavities formed by gas bubbles). In other words, it is vesicular basalt with its holes filled in. Mineral solutions seeped throughout the rock very slowly over the course of millions of years, crystallizing in the gaps and spaces within the basalt. Many amygdaloidal minerals (minerals with cavities filled in by other materials) can be rare and valuable collectibles; examples include agate, pumpellyite, thomsonite, and even copper.

WHERE TO LOOK: Basalt is very common along Lake Superior; look anywhere there is exposed rock.

Basalt, Vesicular

HARDNESS: 5-6 **STREAK:** N/A

Lake Superior Area
Occurrence

AREA: Extremely abundant in all three states

ENVIRONMENT: Lakeshore, gravel pits, mine dumps, and riverbanks

WHAT TO LOOK FOR: Most commonly black, gray, or bluish dense rock with no visible grains

SIZE: Basalt is an extremely common volcanic rock, which means that basalt can be found as huge masses that extend for miles or as tiny pebbles.

COLOR: Basalt can have a great range of color, from dark gray and black to greens, blues, and reds, but is most commonly gray.

OCCURRENCE: Very common

NOTES: Vesicular basalt (basalt which developed holes or cavities) formed at the top of a lava flow, where the gas bubbles within the material had risen. As it cooled, the bubbles were trapped and give the rock a very sponge-like texture. These vesicles (cavities formed by gas bubbles) are very often filled with other minerals that slowly seeped into them and crystallized.

WHERE TO LOOK: Basalt is very common along Lake Superior; look anywhere there is exposed rock.

Copper

Calcite points

Calcite nodule
(round calcite cluster)

Calcite

HARDNESS: 3 **STREAK:** White

Lake Superior Area
Occurrence

AREA: Extremely abundant in all three states

ENVIRONMENT: Lakeshore, gravel pits, mine dumps, and riverbanks

WHAT TO LOOK FOR: White pointed or square crystals when whole, or white nodules (round rock clusters) when found in trap rock (rock that traps other minerals in its gas bubbles).

SIZE: Because it is so common, calcite can be anywhere from pea-sized and smaller all the way up to football-sized specimens.

COLOR: Calcite is most commonly white or clear but is frequently affected by impurities that are stained any color of the rainbow, though yellow is the primary variant.

OCCURRENCE: Very common

NOTES: Calcite is very common; in fact, it is one of the most prevalent minerals on Earth. It commonly fills vesicles (cavities formed by gas bubbles) but also can form large, impressive crystals when allowed the room to grow. In the Lake Superior region, calcite is mostly found as a material which fills in other rocks.

It is very often confused with quartz, but determining one from the other is easily done with a scratch test. Quartz is much harder than calcite. Calcite can easily be scratched with a fingernail, whereas quartz will probably wreck your fingernails. Calcite is very collectible, and good specimens can be quite valuable.

WHERE TO LOOK: Between Duluth and Two Harbors, Minnesota, look along the French River and the Knife River, which run toward Lake Superior. In Michigan, calcite is common in many mine dumps.

Chalcedony

HARDNESS: 7 **STREAK:** White

Lake Superior Area
Occurrence

AREA: Any of the three Lake Superior states

ENVIRONMENT: Lakeshore, gravel pits, riverbeds, and mine dumps

WHAT TO LOOK FOR: Semi-translucent stones in a variety of colors and with conchoidal (curved or circular) fractures

SIZE: Chalcedony can be a wide range of sizes, but is generally softball-sized and smaller.

COLOR: Chalcedony can be a variety of colors, but reds, oranges, whites, greens or browns are most common.

OCCURRENCE: Common

NOTES: Many rock collectors have found a piece of chalcedony and called it an agate, when in reality they're not the same. However, there isn't a lot of difference between the two. Agate is just a banded form of chalcedony, whereas chalcedony occurs more massively and with less definition. It is translucent, and specimens that are thin enough will glow in bright light. Reds and browns are most common, but chalcedony will also often be white or yellow because of the pure quartz within.

Chalcedony is a microcrystalline form of quartz, meaning that it is made up almost entirely of quartz, but the crystals are so small that you can only see them through a microscope. As mentioned before, agate is a banded variety of chalcedony, and jasper is another form, but jasper's crystals are even finer, making the stone more opaque.

WHERE TO LOOK: Chalcedony is found along the lakeshore in all three states. It can also be found in gravel pits, riverbeds, farm fields, and along highways and dirt roads.

Chert

HARDNESS: 7 **STREAK:** White

Lake Superior Area
Occurrence

AREA: Any of the three Lake Superior states

ENVIRONMENT: Lakeshore, gravel pits, riverbeds, and mine dumps

WHAT TO LOOK FOR: Waxy, opaque stones in shades of yellows or grays

SIZE: Chert forms massively and can occur in a wide range of sizes.

COLOR: Yellow, beige, or tan are the primary colors of chert. White, gray, or black specimens are called flint.

OCCURRENCE: Common

NOTES: Chert is a common variety of microcrystalline quartz, akin to chalcedony or jasper. Even under a microscope, quartz grains or crystals are not evident. Chert is a stone that can form massive outcroppings and large beds of stone. It often appears layered and can fool some collectors into thinking they have an agate.

When it is dark-colored, such as black or gray, we call it flint. There is no geological reasoning for this because it is the same mineral. The darker varieties were simply named differently. So while flint is famous for producing a spark when struck, chert, quartz, chalcedony, and even agate will do the same.

Chert, like other minerals related to chalcedony, breaks in a conchoidal, or circular, shape. If you're unsure if what you have is chert, look for circular cracks or chips on the surface.

WHERE TO LOOK: Chert is common in most places around Lake Superior, especially on the shoreline or anywhere there is exposed rock.

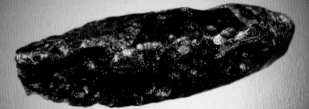

Chlorite nodule
(round chlorite cluster)

Chlorite amygdules
(chlorite which formed in round, gas bubble cavities)

Chlorite

HARDNESS: 2-2.5 **STREAK:** Colorless

Lake Superior Area
Occurrence

AREA: Prevalent in all three states

ENVIRONMENT: Anywhere trap rock (rock which traps other minerals in its gas bubbles) is found

WHAT TO LOOK FOR: Generally dark nodules (round rock clusters) within other rocks

SIZE: Primarily dime-sized or smaller nodules but occurs less frequently as larger, walnut-sized pieces.

COLOR: Dark green to black with a waxy surface luster

OCCURRENCE: Very common

NOTES: Chlorite is an extremely common mineral in the Lake Superior region and is not very sought after or valuable. Chlorite can form in dark bladed crystals, but in this area, it is mostly found as small, dark vesicular nodules or round rock clusters found within the cavities of other rocks, namely basalt and rhyolite. These nodules are generally shades of green, with the most common being so dark that they appear almost black, although some can be rather brightly colored. In Michigan, this can lead one to believe that they have found pumpellyite, a very valuable mineral, but a simple test to determine the difference is to scratch the specimen with a copper coin. Chlorite will easily be scratched by a coin, whereas pumpellyite will not be damaged.

Because chlorite is so common, it can have numerous appearances. The surface of vesicular chlorite (chlorite with cavities) can appear pearly or glassy but commonly it is very dull and plain. And while chlorite crystals are translucent, its nodules are generally opaque.

WHERE TO LOOK: Amygdaloidal basalt and rhyolite often contain chlorite. Look near the lakeshore.

Round stones

Conglomerate

Matrix
(the material other minerals form in)

Breccia

Sharp stones

Conglomerate

HARDNESS: Varies **STREAK:** N/A

Lake Superior Area
Occurrence

AREA: Primarily Michigan but also found in Minnesota and Wisconsin

ENVIRONMENT: Lakeshore and gravel pits

WHAT TO LOOK FOR: Rock that appears to be made up of many smaller round stones, resembling manmade concrete

SIZE: Conglomerate can be found in any size but is generally football-sized and smaller.

COLOR: Can be found in a wide variety of colors but generally is brown or reddish brown with multicolored stones

OCCURRENCE: Common in Michigan, uncommon elsewhere

NOTES: Conglomerate is a unique rock composed of many other smaller rocks. These smaller rocks appear as independent round stones, much like what you can find on the beaches of Lake Superior. The term conglomerate refers to the formation of other minerals between these stones that cement it into a solid rock.

Another type of conglomerate is known as breccia. This forms when a rock, for one reason or another, is crushed into many pieces. Other minerals then come in and cement the rock back together in a random arrangement. The result is very much like a conglomerate, but the stones within are rough, sharp, and jagged.

WHERE TO LOOK: Look any place there is exposed rock, especially along the shore of Lake Superior.

Sheet copper

Copper

HARDNESS: 2.5-3 **STREAK:** Shiny red

Lake Superior Area
Occurrence

AREA: Primarily Michigan, but some is found in Minnesota and Wisconsin

ENVIRONMENT: Lakeshore, gravel pits, and mine dumps

WHAT TO LOOK FOR: Red or greenish blue, heavy, jagged, and malleable pieces of metal

SIZE: Usually found as small flakes or crystals but very rarely as large nuggets or sharp masses

COLOR: While normally reddish, natural copper is almost always oxidized (altered due to exposure to oxygen) to a bright green or blue and is sometimes blackened from great corrosion.

OCCURRENCE: Common

NOTES: Copper is a highly sought after collectible, especially in the Keweenaw Peninsula of Michigan, where a copper-mining industry once thrived. There are many different forms of copper around Lake Superior, but nearly all copper will be corroded to a bluish green color. Ideally, copper is a bright reddish metal, but in nature, rain and air cause it to change colors.

There are three main varieties of copper. Sheet copper is a variety that forms between layers of rock. Float copper comes in nuggets that have been rounded and smoothed by glaciers, and crystal copper is a specimen where the copper has actually grown in its own structure, rather than being influenced by its surroundings. Copper is often associated with calcite, prehnite and, very rarely, agate.

WHERE TO LOOK: While copper is occasionally found on the lakeshore and near riverbanks, it is much more prevalent in the copper mine dumps of Michigan's Keweenaw Peninsula.

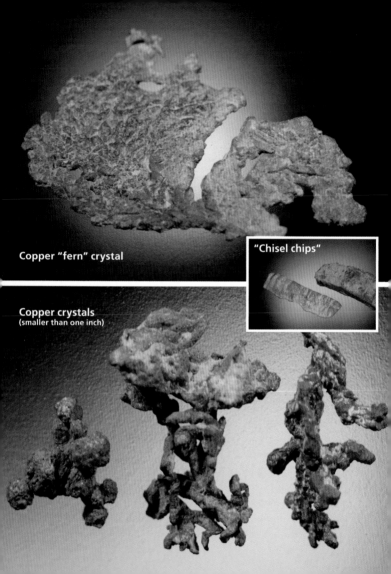

Copper "fern" crystal

"Chisel chips"

Copper crystals
(smaller than one inch)

Copper Varieties

HARDNESS: 2.5-3 **STREAK:** Shiny red

Lake Superior Area
Occurrence

AREA: Primarily Michigan, but some is found in Minnesota and Wisconsin

ENVIRONMENT: Gravel pits and mine dumps

WHAT TO LOOK FOR: Red or greenish blue, heavy, jagged, and malleable pieces of metal

SIZE: Usually found as small flakes or crystals but very rarely as large nuggets or sharp masses

COLOR: While normally reddish, natural copper is almost always a bright green or blue and is sometimes blackened from great corrosion.

OCCURRENCE: Common

There are a great many varieties of copper, both natural and manmade. By far the most desirable and valuable are copper crystals, rare specimens formed when copper solidifies within a cavity large enough for it to take on its true shape rather than just filling the space in a large mass. Copper crystals can be cubic or have points as well as "ferns," which are rare plant-like specimens that grow in a sweeping shape.

A "chisel chip" is manmade. Found primarily in Michigan's Keweenaw Peninsula, chisel chips are made when a miner's chisel chips away at a copper nugget too large for extraction. The miners were after the copper ores, not the sometimes enormous underground nuggets, so they would often go around or chip through large copper masses. Chisel chips are the result of the chisel cutting through the copper, and the width of the chip indicates the width of the chisel used to cut it.

WHERE TO LOOK: While copper is occasionally found on the lakeshore and near riverbanks, it is much more prevalent in the copper mine dumps of Michigan's Keweenaw Peninsula.

Diabase

HARDNESS: 5.5-6 **STREAK:** N/A

Lake Superior Area
Occurrence

AREA: Common in all three states

ENVIRONMENT: Lakeshore and gravel pits

WHAT TO LOOK FOR: Dark to light gray rock with lighter spots throughout that vary in size

SIZE: A volcanic rock, diabase can be found as small pebbles as well as massive boulders.

COLOR: Generally shades of gray but sometimes dark greens and blues. Diabase contains "flowers" of lighter, whiter crystals.

OCCURRENCE: Common

NOTES: Diabase greatly resembles basalt on many occasions but has one major difference: Diabase contains round, white crystals that formed while the rock was cooling. These crystals are often referred to as "flower-like" or "fuzzy" because of the relationship between the white feldspar crystals and the rest of the rock. The crystals formed first, and the darker minerals filled in around them. Because of this process, diabase becomes very hard and dense. The crystals are circular and pale and can be very hard to see in rough specimens but much easier to identify in beach-worn rocks.

Diabase, much like the rest of the volcanic rocks of the Lake Superior region, is very common and has been used as a building material for many years, often crushed and mixed into concrete.

WHERE TO LOOK: Diabase is quite common; look anywhere there is exposed rock.

Microcline feldspar with copper

Feldspar-lined vesicles
(cavities formed by a gas bubble lined with feldspar)

Feldspar

HARDNESS: 6 **STREAK:** White

Lake Superior Area
Occurrence

AREA: Prevalent in all three states

ENVIRONMENT: Lakeshore, gravel pits, mine dumps, and riverbeds

WHAT TO LOOK FOR: Hard orange crystals or masses growing within other rocks

SIZE: As feldspar tends to fill gaps, it can be found as small pea-sized amygdules (feldspar which formed in round, gas bubble cavities) or large masses between other material.

COLOR: Feldspar varies widely in color. Most feldspar around Lake Superior is deep orange or red, but it can also be pink.

OCCURRENCE: Very common

NOTES: Feldspar encompasses a large group of minerals, but Lake Superior has only a few widespread types. Orthoclase feldspar is pink or red and is most commonly found in granite or unakite. Microcline feldspar, however, is more commonly found and is very easy to identify. Microcline feldspar is often found filling vesicles (cavities formed by gas bubbles) and cracks within other rock, generally basalt or rhyolite. It is bright orange, hard, and breaks irregularly, so it shouldn't be very difficult to determine if your specimen is microcline feldspar.

It can often resemble very richly colored specimens of laumontite, but laumontite is very soft when exposed to air and can be scratched with your fingernail, whereas microcline feldspar is much harder.

WHERE TO LOOK: Feldspar is extremely prevalent, especially as amygdules in volcanic rocks like basalt.

Rough specimen

Beach-worn specimen

Gabbro

HARDNESS: 5.5-6 **STREAK:** N/A

Lake Superior Area
Occurrence

AREA: Can be found in all three states but is found primarily in Minnesota

ENVIRONMENT: Lakeshore, gravel pits

WHAT TO LOOK FOR: Dark, mottled rock containing small crystal pieces that are lighter in color

SIZE: Can range greatly in size, from small stones to giant masses

COLOR: Generally dark with lighter spots, sometimes greenish in hue and containing black, reflective crystals

OCCURRENCE: Common in Minnesota, uncommon in Wisconsin and Michigan

NOTES: Gabbro is a very coarse-grained rock that formed like basalt but much deeper within the earth. The volcanic material cooled very slowly, allowing time for larger crystals to form, which is evident by the mottled, multi-color look of gabbro. It often has a greenish hue, and thin specimens can be moderately translucent. Gabbro also contains crystals of black biotite, which results in small reflective spots within the rock.

Gabbro is not as easy to find in Wisconsin and Michigan as it is in Minnesota. Minnesota has one of the country's major formations of the rock, known as the Duluth Gabbro Formation, but it is most easily found farther up Minnesota's North Shore. It can be hard to identify on the beach as the white plagioclase crystals often erode and degrade its distinctive appearance.

WHERE TO LOOK: The Duluth Gabbro Formation, stretching from Duluth, Minnesota to Silver Bay, Minnesota, is the best place to find gabbro on Lake Superior's shore.

Yellow oxidization
(a chemical change due to exposure to oxygen)

Goethite

HARDNESS: 5-5.5 **STREAK:** Yellow to Yellow-brown

Lake Superior Area
Occurrence

AREA: All three Lake Superior states

ENVIRONMENT: Gravel pits and mine dumps

WHAT TO LOOK FOR: Heavy botryoidal (grape-like) or stalactitic (icicle-like) masses of oxidized metal (metal altered due to exposure to oxygen) with a silky texture

SIZE: Goethite can occur in large masses and can be any size but is often found football-sized and larger.

COLOR: Goethite's natural color is a deep gray, but it is almost always oxidized (altered due to exposure to oxygen) to shades of red and yellow.

OCCURRENCE: Common

NOTES: Goethite is an iron ore that is easily found in or near iron deposits. Visually, it can be extremely difficult to distinguish goethite from hematite, which it often occurs alongside. Both minerals oxidize (change due to the exposure of oxygen) to the same colors, so using your eyes simply won't do the job. A streak test is a simple way to determine the difference between the two. Goethite's streak is a yellow or golden yellow brown, whereas hematite's is always reddish brown or rust-colored. When oxidized, goethite turns yellow, while hematite turns red.

Goethite can form in many different ways, but most commonly forms in stalactitic (icicle-like) to botryoidal (grape-like) structures. It has also been found in granular and bladed crystal structures. It feels silky to the touch, and when looking at a broken crystal, it appears fibrous or striated (grooved) inside.

WHERE TO LOOK: Goethite is prevalent in Minnesota's Iron Range around Babbitt, Hibbing, and Virginia, Minnesota, and can also be found in iron-mining areas like Hurley, Wisconsin and Ironwood, Michigan.

Beach-worn specimens

Granitic gneiss

Granite

HARDNESS: 5.5 **STREAK:** N/A

Lake Superior Area
Occurrence

AREA: Found in all three states

ENVIRONMENT: Lakeshore, gravel pits

WHAT TO LOOK FOR: Multicolored, speckled rock containing large grains of light and dark minerals

SIZE: Generally, granite is found on beaches and can range in size from small pebbles to giant boulders.

COLOR: Granite can be a wide range of colors, all of which are mottled with darker spots. Granite is often white, gray, pink, red, or green.

OCCURRENCE: Common

NOTES: Most of the granite found in Minnesota, Wisconsin, and Michigan did not originate there. Granite cooled very slowly in some of the deepest parts of the Earth's crust. The granite on our lakeshores originated in Canada and was pushed here by glaciers.

Granite can be a number of colors, it all depends on what minerals are found in the stone. Pinks generally indicate feldspar, and whites indicate quartz. Granite often contains mica, which is evident as reflective black flecks. Because it cooled so slowly, granite is generally very coarse-grained, and when it has been reheated, it often metamorphoses into a striped variety of granite, known as granitic gneiss.

WHERE TO LOOK: Granite can be found on or near the shore of Lake Superior.

Red oxidization
(a chemical change due to exposure to oxygen)

Botryoidal
(grape-like) surface

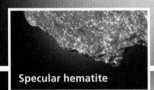

Specular hematite

"Needle ore" hematite

"Needle ore" specimen courtesy of Karl Wanink

Hematite

HARDNESS: 5-6 **STREAK:** Red to Red-brown

Lake Superior Area
Occurrence

AREA: Any of the three Lake Superior states

ENVIRONMENT: Gravel pits and mine dumps

WHAT TO LOOK FOR: Dark, metallic, iron-like material often with a sparkling or glittering luster

SIZE: Hematite occurs in a wide range of sizes, from massive pieces to small specimens.

COLOR: Black metallic colors are ideal, but as an iron ore, it is often oxidized (exposed to oxygen) and erodes to red or orange rust-like colors.

OCCURRENCE: Common

NOTES: Hematite is the most common iron ore and occurs in many different forms. There are massive sedimentary deposits in Northern Minnesota, Wisconsin, and Michigan, and it has been a leading source of industry in these areas for many years. It is highly collectible, and unique specimens can be worth quite a bit of money. Hematite forms most commonly in botryoidal (grape-like) masses but can also be radiating (spread out from a central point) or stalactitic (icicle-like). A variety known as specular hematite is also highly collected, as it appears to sparkle like silver.

Another form of iron ore, called goethite, is often confused for hematite, and in fact, there is very little difference between the two. The most important difference to know is the streak color. Hematite's streak is rust-red, while goethite's is always a golden yellow.

WHERE TO LOOK: Like goethite, hematite is prevalent in Minnesota's Iron Range around Babbitt, Hibbing, and Virginia, Minnesota, and can also be found in iron-mining areas like Hurley, Wisconsin and Ironwood, Michigan.

Beach-worn specimens

"Jaspelite" — banded iron ore and jasper

Jasper

HARDNESS: 7 **STREAK:** White

Lake Superior Area
Occurrence

AREA: Extremely prevalent in all three states

ENVIRONMENT: Very common in virtually every location

WHAT TO LOOK FOR: Dark red opaque rock with a glassy or waxy luster

SIZE: Jasper has been found in a wide range of sizes, from pea-sized samples up to basketball-sized and larger stones.

COLOR: Deep reds and purples are most common, but it can also be green or yellow.

OCCURRENCE: Very common

NOTES: Jasper is one of the Lake Superior area's most prominent and collectible stones. It is one of the many forms of microcrystalline quartz, making it very closely related to chert, chalcedony, and agate. It is generally a deep red or purple that is very smooth and waxy to the touch. Some jasper can appear banded, leading many people to believe that what they've found is an agate. The primary difference between the two is that jasper's crystal particles are much smaller than agate's, making it totally opaque in most situations, whereas agate is translucent, and some specimens will appear to glow under very bright light.

A variety of jasper banded with iron ore is unofficially called "jaspelite." This particular type of jasper formed billions of years ago in the Earth's ancient oceans when sediment formed into bands as it solidified.

WHERE TO LOOK: Jasper is easy to find around Lake Superior, especially on the shore and any place there is exposed rock.

Driftwood

Porcelain tile

Slag glass

Beach glass

Rusty metal

Aluminum "blob"

Concrete

Brick

Blacktop

Coal

Junk

HARDNESS: Varies **STREAK:** N/A

AREA: Any of the three Lake Superior states

Lake Superior Area
Occurrence

ENVIRONMENT: Lakeshore

WHAT TO LOOK FOR: Anything that looks out of place, namely pieces of metal, glass, wood, and any stones different from those around them

SIZE: Beach junk can be as small as a shard of glass or as large as furniture or appliances.

COLOR: Naturally, the color of a piece of beach junk varies greatly depending on what it is.

OCCURRENCE: Very common

NOTES: Most beach junk is very easy to find and identify, with chunks of metal, plastic, and wood being the most obvious. However, not everything is so easily classified as "junk." Some beach glass can closely resemble quartz, but unlike quartz, glass is softer and can be scratched with a steel tool, such as a file. Once beach-tumbled, tile, porcelain, and brick can look extremely similar to some rocks, such as limestone. Other things end up on Lake Superior's shores as a result of the area's past and present industries. Concrete, tar, coal, taconite pellets, aluminum "blobs," and slag glass are just some of the items dumped into the lake by factories and businesses, either intentionally or accidentally. The best way to determine the difference between a rock and a look-alike piece of junk is to learn the traits of the rock in question. Familiarize yourself with its characteristics, such as streak and hardness.

WHERE TO LOOK: You can find beach junk all around the shores of Lake Superior.

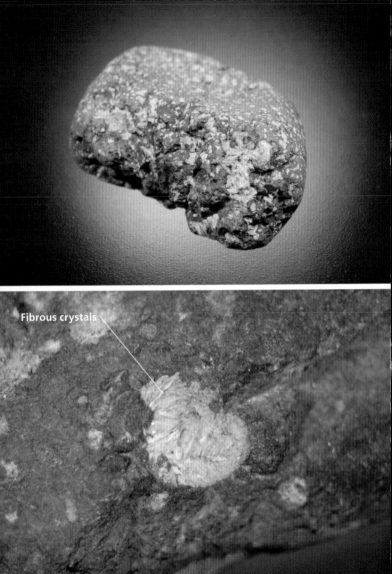

Fibrous crystals

Laumontite

HARDNESS: 4 **STREAK:** Colorless

Lake Superior Area
Occurrence

AREA: Any of the three Lake Superior states

ENVIRONMENT: Lakeshore, gravel pits, riverbeds, and mine dumps

WHAT TO LOOK FOR: Light-colored, fibrous, and very soft crystals within vesicles (cavities formed by gas bubbles) of other rock

SIZE: Laumontite is most commonly found as an amygdule (a mineral within another rock), making it normally no larger than a coin in size.

COLOR: Laumontite is white in its purest form but almost always includes inclusions (minerals that formed within it) that make it pink or orange.

OCCURRENCE: Very common

NOTES: Laumontite is Lake Superior's most common member of the zeolite mineral family. Zeolites are minerals with radiating crystals or a fan-like structure. Laumontite is very fibrous and crystalline and has a pearly luster and can sometimes appear translucent. The primary characteristic that separates it from other zeolites is its brittleness. When laumontite is exposed to air, it rapidly dehydrates and becomes extremely brittle and soft. Because of this, many specimens are heavily weathered, and all crystal structure is destroyed. Very orange laumontite can resemble microcline feldspar, another common vesicle-filling (cavity-filling) mineral, but laumontite is much softer and can be identified with a hardness test. Exposed laumontite should be damaged by a fingernail while the feldspar should not.

WHERE TO LOOK: Laumontite often occurs as amygdules in volcanic rocks like basalt, and can be found all around Lake Superior.

Granular magnetite

Specimen courtesy of A.E. Seaman Mineral Museum

Magnetite layered with quartz

Specimen courtesy of A.E. Seaman Mineral Museum

Magnetite

HARDNESS: 5.5-6.5 **STREAK:** Black

Lake Superior Area
Occurrence

AREA: Any of the three Lake Superior states

ENVIRONMENT: Gravel pits and mine dumps

WHAT TO LOOK FOR: Heavy, dark iron material that exhibits many traits of hematite but is magnetic

SIZE: Magnetite can occur massively and in very large masses.

COLOR: Magnetite is metallic and black in color, much like hematite.

OCCURRENCE: Uncommon

NOTES: Magnetite is a variety of iron ore found in places where iron is characteristically mined. It is extremely similar to hematite and, in fact, differs very little from it. To identify magnetite, simply attach a magnet to the stone. If it sticks, it is magnetite. Another test is the streak method: Hematite streaks reddish brown while magnetite's streak is a dark gray or black color.

Magnetite can form in many different ways, much like hematite or goethite. It can form with bands of quartz or jasper as well as be stalactitic (icicle-like) and granular. It can have a metallic luster, but this is not always present. In rare cases, magnetite can be found constituted within Mary Ellen jasper. This is not common and has only been discovered in the Iron Range of Minnesota.

Magnetite is not particularly valuable or collectible except in rare specimens. But thanks to a simple magnet, it is one of the easier minerals to identify.

WHERE TO LOOK: Like other minerals found alongside iron ores, magnetite is prevalent in Minnesota's Iron Range around Babbitt, Hibbing, and Virginia, Minnesota, and can also be found in iron-mining areas like Hurley, Wisconsin and Ironwood, Michigan.

Foliated (sheet-like) layers

Mica

HARDNESS: 2-3 **STREAK:** N/A

Lake Superior Area
Occurrence

AREA: Any of the three Lake Superior states

ENVIRONMENT: Lakeshore, gravel pits, riverbeds, and mine dumps

WHAT TO LOOK FOR: Dark, layered, reflective crystals on or in other rocks

SIZE: Mica is generally found as small crystals or flecks within other rock.

COLOR: There are many colors of mica, but it is more commonly found as dark, brown to black, slightly metallic crystals.

OCCURRENCE: Common

NOTES: The term mica refers to a group of minerals, rather than just one particular stone. There are many different varieties of mica in the Lake Superior region, including muscovite, phlogopite, biotite, and chlorite. These minerals are easily found but not often as individual specimens. Most micas you'll find are a part of another rock, because mica is a very common rock-building mineral. Granite often has a lot of mica within it, visible as small, dark, reflective flecks of color. Mica generally occurs in stacks of paper-thin sections of the mineral, and it is very soft and can be scratched or broken with your fingernail.

WHERE TO LOOK: Mica is most often found in combination with other rocks, especially granite.

Basalt porphyry

Long, jagged crystals

Porphyry

HARDNESS: Varies **STREAK:** N/A

Lake Superior Area
Occurrence

AREA: Any of the three Lake Superior states

ENVIRONMENT: Lakeshore, gravel pits, riverbeds

WHAT TO LOOK FOR: Finely grained volcanic rock containing larger irregular crystals

SIZE: Porphyry can be found in a dramatic range of sizes.

COLOR: As there are different types of porphyry, it varies greatly. Generally it is gray rock containing crystals of varying colors, such as red, yellow, or brown.

OCCURRENCE: Common

NOTES: Porphyry is actually a general term for volcanic rock that contains larger crystals of other minerals. Commonly, this means basalt and rhyolite, but less often, you can find granite porphyry.

Many people confuse amygdaloidal basalt (basalt with its cavities filled in) with porphyry, but it's easy to tell the two apart once you know what you're looking for. The crystals in porphyry formed while the rock was cooling; therefore the minerals within the stone are more irregular and sharply shaped. Amygdaloidal basalt (basalt with cavities filled in with another mineral), formed when the basalt near the top of a lava flow cooled with many air bubbles within. These air bubbles, over time, filled in with other minerals. So the main difference between porphyry and amygdules is crystal shape: Porphyry is rough, jagged, and generally has long crystals, while basalt amygdules are very round and smooth.

WHERE TO LOOK: Porphyry is often found near exposed rock. It is commonly found in basalt, rhyolite, and granite.

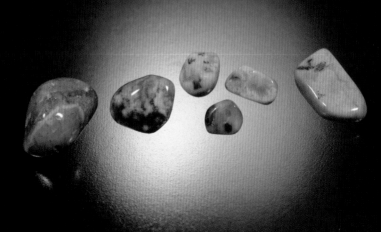

Copper inclusions
(copper which formed within the prehnite)

Botryoidal
(grape-like) surface

Polished pieces

Prehnite

HARDNESS: 6-6.5 **STREAK:** Colorless

Lake Superior Area
Occurrence

AREA: All three states

ENVIRONMENT: Gravel pits, mine dumps and, less commonly, the lakeshore

WHAT TO LOOK FOR: Pale green crystals within rock cavities or frosted green nuggets when found on the shore

SIZE: Prehnite generally occurs in pieces from peanut-size up to very large masses or layers.

COLOR: Prehnite is pale green most of the time, but its coloration can be very faint, making it almost white or gray in color.

OCCURRENCE: Uncommon

NOTES: Prehnite is a collectible stone that is rather widespread throughout the Lake Superior area, but some of the best specimens come from the Keweenaw Peninsula of Michigan. It forms in botryoidal (grape-like) masses generally within cavities in other rock. Normally, it is a pale green color, but less impressive specimens will be white or gray.

Prehnite can often be found occurring with other minerals, namely calcite, epidote, various zeolites, and copper. Prehnite grows around other materials and can actually encompass them completely. Good examples of this have been found in Michigan specimens of prehnite, where the mineral appears to have copper within it. The prehnite actually formed after the copper and grew right around it.

WHERE TO LOOK: Prehnite occurs in cavities of volcanic rock and can be found on riversides near Minnesota's North Shore from Duluth to Silver Bay, Minnesota. In Michigan, look in the mine dumps of the Keweenaw Peninsula.

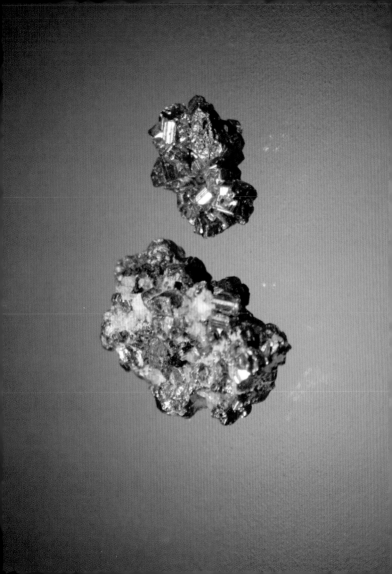

Pyrite

HARDNESS: 6-6.5 **STREAK:** Greenish Black

Lake Superior Area
Occurrence

AREA: Minnesota, Wisconsin, and Michigan

ENVIRONMENT: Gravel pits and mine dumps

WHAT TO LOOK FOR: Golden metal with definite crystal structures.

SIZE: Pyrite is generally found as small crystals no larger than a dime or in large masses.

COLOR: Pyrite crystals are golden and metallic.

OCCURRENCE: Common

NOTES: Pyrite is a common iron-based mineral that has been referred to for years as "fool's gold" because of its color's great resemblance to gold. It generally forms on top of other minerals or in cavities in rock, growing in a wide array of crystal structures. Most commonly, pyrite is cubic in structure but can also be octahedral (having eight plane faces) or even botryoidal (grape-like structure). Perfectly cubic or pyramid-shaped crystals are the most sought after.

Pyrite is extremely collectible and large, well-formed specimens can be very valuable. It is often found growing as a thin coating on quartz and other semi-precious minerals, which can make for unique display specimens. Pyrite is found all over the world in many shapes and sizes, but in the Lake Superior region, it generally remains small and appears most common as small, broken fragments. As you can imagine, pyrite's distinct color and bright metallic look can make it easy to spot.

WHERE TO LOOK: Pyrite, like other iron-based minerals, is prevalent in Minnesota's Iron Range around Babbitt, Hibbing, and Virginia, Minnesota, and can also be found in iron-mining areas like Hurley, Wisconsin and Ironwood, Michigan.

Pyrolusite

HARDNESS: 6-6.5 **STREAK:** Black to Blue-black

Lake Superior Area
Occurrence

AREA: Minnesota, Wisconsin, and Michigan

ENVIRONMENT: Gravel pits and mine dumps

WHAT TO LOOK FOR: Fibrous, fan-shaped crystals with a metallic luster

SIZE: Pyrolusite is generally small, with crystals being no bigger than your thumb.

COLOR: Silver-black crystals with a dull metallic luster.

OCCURRENCE: Uncommon

NOTES: Pyrolusite is a small, fibrous crystal composed of oxidized manganese (manganese exposed to oxygen). It is silvery black or gray with crystals that radiate outward from a central point in a fan-like shape. It is commonly found growing on the surface of other rocks and can form very complex angular shapes. Generally, it is not particularly valuable with the exception of very well-formed specimens. Some people have mistaken its plant-like shape for fossils. Industrially, pyrolusite is mined and used as an important source of manganese.

Pyrolusite greatly resembles manganite, another dark, metallic, fan-shaped mineral, and it is very hard to differentiate between the two with the naked eye. The most easily identifiable difference is the streak color. While manganite's streak is dark reddish brown, pyrolusite's streak has a bluish tint to it. Pyrolusite can be found wherever earth is being moved, namely gravel pits, or less commonly, mine dumps.

WHERE TO LOOK: Pyrolusite, while rarer than other iron-based minerals, can be found in Minnesota's Iron Range around Babbitt, Hibbing, and Virginia, Minnesota, and can also be found in iron-mining areas like Hurley, Wisconsin and Ironwood, Michigan.

Beach-worn specimens

Quartz-lined vesicles
(quartz-lined cavities formed by a gas bubble)

Quartz

HARDNESS: 7 **STREAK:** White

Lake Superior Area
Occurrence

AREA: Extremely prevalent in all three states

ENVIRONMENT: Lakeshore, riverbeds, gravel pits, and mine dumps – virtually everywhere

WHAT TO LOOK FOR: White crystals or masses that often fill in the holes or gaps in other rock, or as white, rounded beach pebbles on the lakeshore

SIZE: Quartz is very common and can be found in a wide range of sizes, but most commonly, it is walnut-sized or smaller when found on the beach.

COLOR: Quartz is generally white or clear but can be stained or tinted to a different color, including gray, yellow, purple, pink or red.

OCCURRENCE: Very common

NOTES: Quartz is one of the most common minerals in the world and can be found in almost every geological environment. It plays a very large role in forming many rocks and minerals. In the Lake Superior region, it is easily found as white, rounded beach pebbles, or within vesicles (cavities formed by gas bubbles) with a thin lining of crystal points. You can commonly find it filling cracks in large basalt flows.

Quartz is responsible for many of Lake Superior's collectible stones, such as agate and jasper. In fact, these stones are almost made up entirely of hard, dense quartz.

WHERE TO LOOK: Quartz is very common around the lake, and it is particularly easy to find where there is exposed rock, especially near the lakeshore.

Quartz on Chert

HARDNESS: 7 **STREAK:** White

Lake Superior Area
Occurrence

AREA: Extremely prevalent in all three states

ENVIRONMENT: Lakeshore

WHAT TO LOOK FOR: White crystals or masses that often fill in the holes or gaps in other rock, or as white, rounded beach pebbles on the lakeshore

SIZE: Quartz is very common and can be found in a wide range of sizes, but most commonly, it is walnut-sized or smaller when found on the beach.

COLOR: Quartz is generally white or clear but can be stained or tinted to a different color, including gray, yellow, purple, pink or red.

OCCURRENCE: Common

NOTES: Quartz is very often found with other minerals, such as jasper, chert, chalcedony, and agate. In addition, it is also a secondary mineral that fills in cracks, gas bubbles, and faults in rock long after they've formed. Agates are a common and popular example. Quartz can also fuse a rock back together by filling in a crack that had once broken the stone in two.

WHERE TO LOOK: Quartz is very common around the lake, as is chert. Look where there is exposed rock, especially near the lakeshore.

Quartz, Iron-stained

HARDNESS: 7 **STREAK:** White

Lake Superior Area
Occurrence

AREA: Extremely prevalent in all three states

ENVIRONMENT: Mine dumps and overburden piles

WHAT TO LOOK FOR: White crystals or masses that often fill in the holes or gaps in other rock, or as white, rounded beach pebbles on the lakeshore

SIZE: Quartz is very common and can be found in a wide range of sizes, but most commonly, it is walnut-sized or smaller when found on the beach.

COLOR: Quartz is generally white or clear but can be stained or tinted to a different color, including gray, yellow, purple, pink or red.

OCCURRENCE: Common

NOTES : Quartz can vary greatly in color based on its impurities or staining. Amethyst is a purple variety of quartz, rose quartz is a pink variety and smoky quartz is a gray variety. The color depends on the surroundings. For instance, most specimens of red quartz are found in or around iron mines, and the red coloring is actually the result of oxidization (a chemical change due to exposure to oxygen).

WHERE TO LOOK: Look near iron mines, especially in mine dumps and overburden piles.

Quartzite

HARDNESS: Varies **STREAK:** N/A

Lake Superior Area
Occurrence

AREA: Minnesota, Michigan, and Wisconsin

ENVIRONMENT: Lakeshore and gravel pits

WHAT TO LOOK FOR: Very hard, light-colored stones greatly resembling quartz but having more texture

SIZE: Quartzite is mostly found as small, fist-sized or smaller stones.

COLOR: Quartzite can be a wide range of slightly translucent colors, most of which are light, including white, cream, and yellow.

OCCURRENCE: Common

NOTES: When sandstone, which consists of mostly silica (quartz), is subjected to great pressure, it turns into quartzite. Quartzite is very similar to quartz in that it is extremely hard and is resistant to weathering, but it has some key differences from quartz. Since quartzite was originally sandstone, it has a slightly grainy appearance, whereas quartz does not. Quartzite can be white but is generally yellow or other colors, depending on impurities within the stone.

Sandstone is loosely cemented together, and you can separate grains of sand from its surface, but once it has been metamorphosed into quartzite, it becomes as hard and impenetrable as quartz. When quartzite is broken, it breaks through its grains, while sandstone breaks around its grains.

WHERE TO LOOK: Quartzite can be most easily found on beaches, especially on the eastern end of Michigan's Upper Peninsula.

Beach-worn specimens

Amygdaloidal rhyolite
(rhyolite with its cavities filled in)

Rhyolite

HARDNESS: 6-6.5   **STREAK:** N/A

Lake Superior Area
Occurrence

AREA: Prevalent in all three states

ENVIRONMENT: Lakeshore, gravel pits, riverbeds, and mine dumps

WHAT TO LOOK FOR: Red, grainy rock, often with gas bubbles, sometimes so many that the rock appears spongy

SIZE: Rhyolite is a volcanic rock and therefore can be found in any size, from pebbles to entire cliffs.

COLOR: Most commonly occurs in dark reds, but is sometimes lighter and more orange. Rhyolite can also be found as a more purple stone.

OCCURRENCE: Very common

NOTES: Rhyolite is another extremely common volcanic rock that resulted from the numerous lava flows that once occurred in the Lake Superior area. Like basalt, it can be found around the lake and even shares some of basalt's colors, but the biggest difference between the two is the mineralogical consistency of the stones. While basalt cooled so quickly that no grains can be seen even under a microscope, rhyolite actually looks a bit like red granite under magnification. In fact, rhyolite is virtually the same mineralogical composition as granite, and if it had cooled slower, it would have become granite.

WHERE TO LOOK: Look for rhyolite anywhere there is exposed rock.

Sandstone

HARDNESS: Varies **STREAK:** N/A

Lake Superior Area
Occurrence

AREA: All three states, but primarily
Upper Michigan

ENVIRONMENT: Lakeshore, gravel pits, riverbeds

WHAT TO LOOK FOR: Light-colored, coarsely grained rock
resembling the texture of sandpaper

SIZE: Sandstone can be massive but can be found in any size.

COLOR: Red, yellow, and brown are the most common colors
of sandstone.

OCCURRENCE: Common, very common in Michigan

NOTES: Sandstone is the result of sand compacting and solidifying over many years. It can easily be scratched and even pulled
apart with your hands, though you probably wouldn't want to;
the sand grains, mostly quartz, are very hard and sharp.

Sandstone normally occurs in shades of brown and red but
occasionally appears in yellow and beige. It can exhibit different
patterns, such as spots and swirls of lighter or darker colors,
which are simply a result of different colors of sand or impurities
in the stone. While sandstone is easily found in Upper Michigan,
it is far less common in Minnesota and Wisconsin, and any specimens found in those two states likely came from elsewhere. If
the cement holding the grains together is also quartz, the stone
can undergo a metamorphosis and become a much harder stone,
known as quartzite.

WHERE TO LOOK: Sandstone can be found anywhere along the
lakeshore, but it is primarily found on Michigan's beaches east of
the Keweenaw Peninsula.

Shale

HARDNESS: Varies **STREAK:** N/A

Lake Superior Area
Occurrence

AREA: All three states

ENVIRONMENT: Gravel pits and mine dumps

WHAT TO LOOK FOR: Light-colored, layered rock that appears as large, flat plates of rock broken into irregular shapes

SIZE: Shale can be found in any size, even as entire cliffs.

COLOR: Shale is mostly dark to light gray but can also be tan or brown, based on the minerals it forms in.

OCCURRENCE: Common

NOTES: Shale, like sandstone and siltstone, is a result of sedimentary minerals settling and solidifying to form a solid rock. It is finer-grained than sandstone but more coarse than siltstone and is often found in layered sheets of rock. Due to their similar natures, shale often occurs with siltstone, and there are even some localities where the two minerals form entire cliffs or outcroppings together. Shale weathers easily and quickly on the beaches of Lake Superior, and the layers within the rock can be more easily seen in highly eroded specimens.

Shale often contains the remains of organic material and therefore can contain fossils, such as clam or snail shells, but rarely more delicate things, such as leaves. And when shale is subjected to heat and pressure, it metamorphoses into the dark and very brittle mineral, slate. Slate is very uncommon in the Lake Superior region.

WHERE TO LOOK: Shale can be found inland, especially where there is exposed rock.

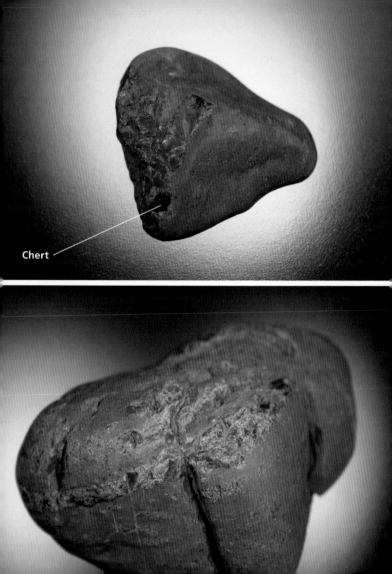

Chert

Siltstone

HARDNESS: Varies **STREAK:** N/A

AREA: All three states

ENVIRONMENT: Lakeshore

WHAT TO LOOK FOR: Extremely fine-grained and dense rock weathered with a very smooth surface texture

SIZE: Siltstone is generally found in pieces no larger than a softball.

COLOR: Dark brown to yellow brown, many times with dark gray or black fragments within.

OCCURRENCE: Common

NOTES: Siltstone, like shale or sandstone, is the result of sediment solidifying into a solid stone. In this case, the finest of all sediments, silt and clay, solidified and formed siltstone. This rock can most often be found as weathered stones on the lakeshore and is very finely grained. Because of this, siltstone is extremely smooth-textured and one of the smoothest rocks you can find on the beach.

One of the most common occurrences in siltstone is the inclusion of dark gray chert. These hard, dense pieces within the yellow brown siltstone are often found as bands or stripes throughout the stone. Since the chert is much harder than the siltstone, it weathers more slowly and the bands often feel raised above the surface of the rest of the stone. Siltstone is easy to find and just as easy to identify, unlike some of the rocks found on Lake Superior's shores.

WHERE TO LOOK: Siltstone is prevalent on many of Lake Superior's beaches.

Fibrous crystal structure

Zeolite

HARDNESS: 3.5-5.5 **STREAK:** Colorless

Lake Superior Area
Occurrence

AREA: Found in all three states

ENVIRONMENT: Anywhere trap rock (or rock that traps other minerals in its gas bubbles) can be found

WHAT TO LOOK FOR: Pale, fibrous crystal nodules (round rock clusters)

SIZE: Lake Superior's zeolites are mostly small nodules (round rock clusters), no bigger than a penny in size.

COLOR: Zeolites are generally pale pink to white but can be reddish, orange or pale green, based on the impurities.

OCCURRENCE: Common

NOTES: The term zeolite actually refers to a group of minerals that all exhibit some of the same properties. Generally, they all appear fibrous in nature and have a radiating structure. Most are white, but some are pink or reddish. They form in the cavities of volcanic rocks, namely basalt, as a reaction of alkaline water (water with a pH-level higher than seven) mixing with volcanic ash. They take millions of years to crystallize and harden, but zeolites are much softer than quartz-based minerals.

Zeolites in the Lake Superior region include analcime, mesolite, and stilbite, among others. The orange laumontite is one of the most common zeolites, and thomsonite is one of the most valuable and rare. Thomsonite is found primarily on Minnesota's North Shore, up near the Canadian border, and is highly sought after. The best specimens of thomsonite have green and pink "eyes" and the characteristic radiating crystal fibers.

WHERE TO LOOK: Zeolites are found all over the Lake Superior region as amygdules in volcanic rocks such as basalt.

Minnesota

Split Rock Lighthouse, north of Two Harbors, sits high atop a cliff of Minnesota anorthosite.

Agates are Minnesota's state gemstone for a reason.

More Lake Superior agates are found in Minnesota than in any other state. Minnesota is also home to thomsonite, one of Lake Superior's most valuable gemstones as well as vast, beautiful beaches full of collectible rocks and minerals all easily accessible with Highway 61 following Lake Superior's beautiful shoreline. Duluth, Two Harbors, Beaver Bay, Silver Bay, and Grand Marais are all locations not to be missed for the serious collector. With a rich iron-mining history and one of Lake Superior's most prominent shorelines, Minnesota is a rock picker's dream.

Rough specimens

Polished specimens

Agate, Paradise Beach

HARDNESS: 7 **STREAK:** White

Lake Superior Area
Occurrence

AREA: Minnesota

ENVIRONMENT: Lakeshore, gravel pits, mine dumps, and riverbanks.

WHAT TO LOOK FOR: Translucent, banded stones of varying colors, often with quartz crystals.

SIZE: Lake Superior agates can be any size, from tiny grains of sand up to softball-sized specimens

COLOR: Orange and browns, but also pink, red, white, or blue

OCCURRENCE: Rare

NOTES: Paradise beach agate is a variety of paint agate that is richly colored in oranges and browns, named for the beach where it was first found. They can also be pink, red, white, or blue.

Paradise Beach agates are quite rare and are highly collectable. Near the beach, they can still be found in the host rock, but attempting to chisel them free is prohibited. Most of the agates found there now are small nodules (round rock clusters) with a black husk. This beach is very accessible to the highway and is very heavily picked. One of the finest collections of Paradise Beach agate is on display at Agate City Rocks and Gifts in Two Harbors, Minnesota.

WHERE TO LOOK: Paradise Beach is located about 12.5 miles north of Grand Marais, Minnesota; however, much of the beach is privately owned and picking is prohibited on private property.

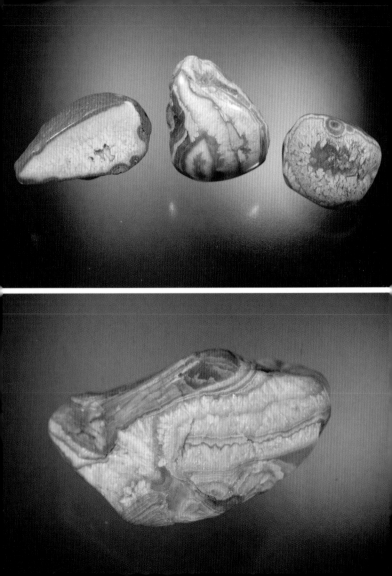

Agate, Skip-an-Atom

HARDNESS: 7 **STREAK:** White

Lake Superior Area
Occurrence

AREA: Minnesota

ENVIRONMENT: Lakeshore, gravel pits, mine dumps, and riverbanks.

WHAT TO LOOK FOR: Translucent, banded stones of varying colors, often with quartz crystals.

SIZE: Lake Superior agates can be any size, from tiny grains of sand up to softball-sized specimens

COLOR: Agates can run the full range of colors, but the most common colors are red, brown, gray, or white with clear, translucent quartz

OCCURRENCE: Rare

NOTES: These agates are exclusive to Minnesota's shoreline and are one of the most unique types in the area. They tend to exhibit little banding and instead have jagged arrangements of bluish white opaque quartz. The peculiar name results from some people's belief that this strange crystal structure is formed by a "missing" atom in the water molecules, though the real reason hasn't been determined.

WHERE TO LOOK: These rare agates can be found from Duluth to Two Harbors, Minnesota, and farther up the shore to Silver Bay, Minnesota. Flood Bay in Two Harbors, Minnesota is a good collecting site.

Crystal points

Red iron staining

Beach-worn specimens

Specimen courtesy of David Gredzens

Amethyst

HARDNESS: 7 **STREAK:** None

AREA: Primarily Minnesota

Lake Superior Area
Occurrence

ENVIRONMENT: Gravel pits, lakeshore, and mines

WHAT TO LOOK FOR: Crystal points or nodules (round rock clusters) resembling quartz, tinted purple or red

SIZE: Beach amethyst generally is found as nodules (round rock clusters) no bigger than a walnut, but good crystal specimens can be very large.

COLOR: Pale to deep purples are the most sought after, but reds and whites are common varieties that are less desirable.

OCCURRENCE: Uncommon

NOTES: Amethyst is a purple variety of quartz that has been used for years in jewelry and as a collectible gemstone. The deepest purple specimens are the most valuable, but the majority are pale purple to white. Lake Superior amethyst, typically coming from Canada, is often tinted red, due to iron impurities.

Large amethyst points can be very valuable, but generally, they form as a coating of small crystals on or in another material. It is uncommon, but amethyst can be found on the beach as rounded pebbles. Most beach specimens have been worn down to the point of losing all of the purple color and simply look like quartz. The best specimens are found across the border in Canada, but it can be found in Minnesota and even inside of agates.

WHERE TO LOOK: Amethyst can be found on the shoreline from Two Harbors, Minnesota to the Canadian border; it is more common close to the border.

Anorthosite

HARDNESS: 5.5 **STREAK:** N/A

Lake Superior Area
Occurrence

AREA: Minnesota

ENVIRONMENT: Lakeshore

WHAT TO LOOK FOR: Light-colored massive rock with granular texture and slightly translucent crystals

SIZE: Anorthosite can range in size from small stones up to entire cliffs.

COLOR: Anorthosite is white to light gray, often with greenish or yellow translucent crystals.

OCCURRENCE: Uncommon

NOTES: Anorthosite is a unique rock in that it is made almost entirely from one mineral: plagioclase feldspar. There is nearly no quartz in the rock, and what color it has is attributed to inclusions (minerals such as olivine and pyroxene that formed within it).

Anorthosite is closely associated with gabbro and can occur together in layers. One of Minnesota's most notable landmarks, Split Rock Lighthouse, is situated on top of a cliff that is composed entirely of anorthosite. In fact, it is one of the world's largest visible masses of anorthosite. And another unique fact is that parts of the moon's surface have been found to be formed of anorthosite.

Anorthosite has also played a part in Minnesota's history when the 3M Corporation, founded in Two Harbors, just north of Duluth, built a mine on the lakeshore, thinking the anorthosite was corundum, a chief material used in sandpaper.

WHERE TO LOOK: Two good places to find anorthosite are around Split Rock Lighthouse about 20 miles northeast of Two Harbors, Minnesota and Illjen City, about five miles northeast of Silver Bay, Minnesota.

Rough specimen

Polished slab

Binghamite

HARDNESS: Varies **STREAK:** N/A

Lake Superior Area
Occurrence

AREA: Minnesota's Iron Range

ENVIRONMENT: Gravel pits and mines

WHAT TO LOOK FOR: Colorful, fibrous material with quartz and iron-like areas

SIZE: Binghamite's size varies greatly, but it is generally no larger than football-sized chunks.

COLOR: Binghamite can have a wide array of intense colors, from yellows to oranges and reds, with black or silver black iron inclusions (iron which formed within it).

OCCURRENCE: Rare

NOTES: Binghamite is one of Minnesota's more rare and unique stones. Its beautiful colors and sweeping fibers make it a highly collectible stone. Binghamite forms when quartz begins to replace hematite, goethite, and other iron ores, creating long, fibrous flows of color often intermixed with bands of black iron. Pure-white quartz can also fill in cavities within the stone.

There is little difference between binghamite and silkstone, another one of Minnesota's Iron Range minerals. The two minerals form together in the same specimen and were considered the same material until they were thoroughly studied. Binghamite is identified best by its long banded fibers, whereas silkstone generally has shorter, more irregular, and less organized fibers.

WHERE TO LOOK: Binghamite can be found on Minnesota's Iron Range, especially near Crosby, Minnesota.

Rough specimen

Cut specimen

Quartz inclusion
(quartz which formed within the greenstone)

Ely Greenstone

HARDNESS: Varies **STREAK:** N/A

Lake Superior Area
Occurrence

AREA: Northern Minnesota

ENVIRONMENT: Gravel pits

WHAT TO LOOK FOR: Deep green, dense, heavy rock

SIZE: Formations of the rock can be massive, but generally you can find pieces bowling ball-sized and smaller.

COLOR: Dark to light green with white quartz inclusions (quartz that formed within it).

OCCURRENCE: Uncommon

NOTES: Ely greenstone is an incredibly old mineral, dating back 2.6 billion years, making it one of the most ancient materials in the world. The Ely greenstone belt is approximately 40 miles long and ranges from 2 to 6 miles wide. This is the one of the only places in the world that this particular variety of rock can be found. It can be collected or purchased in the Ely area in northern Minnesota.

This greenstone formed when underwater volcanic basalt flows cooled and were subjected to great heat and pressure. This resulted in a metamorphosis of the basalt and the minerals affecting it, such as epidote, chlorite, and actinolite, which changed the color to the deep, rich green the stone is named for.

The stone itself isn't particularly valuable, but it is rather collectible because once cut and polished, you can clearly see the unique structure of the rock. The white or gray quartz inclusions (quartz that formed within it) also add to the stone's character.

WHERE TO LOOK: Ely Greenstone is found in and around Ely, Minnesota, which is about 70 miles north of Two Harbors, Minnesota.

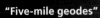
"Five-mile geodes"

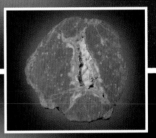

"Eveleth geodes"

Geodes

HARDNESS: Varies **STREAK:** N/A

AREA: Northern Minnesota

Lake Superior Area
Occurrence

ENVIRONMENT: Gravel pits and lakeshore

WHAT TO LOOK FOR: Round balls of rock resembling the surrounding rocks but with crystals or a cavity inside

SIZE: Geodes tend to be baseball-sized and smaller.

COLOR: The outer appearance of geodes is that of surrounding rock, so it can be reddish, brown, or gray.

OCCURRENCE: Uncommon

NOTES: Geodes are not common in most localities around the Lake Superior area, but there are a few types in Minnesota that can be found. The first are known as "five-mile geodes" and are found just north of Grand Marais on Lake Superior's North Shore. These geodes look like small, hard, round clay balls that are very indistinct on the outside, but on the inside, they generally have a thin segment of quartz or chalcedony often surrounding a small cavity.

Another type of Minnesota geodes are called "Eveleth geodes," named after the nearby town. These are small, round balls of rock whose outsides greatly resemble that of Wisconsin's concretions. Unlike concretions, Eveleth geodes have crystals inside them. When hunting these geodes, look near the iron mines in the area and watch for nearly perfectly spherical yellow rocks with a powdery golden oxide on the surface that easily rubs off on your hands.

WHERE TO LOOK: "Eveleth geodes" can be found in the iron mine dumps near Eveleth, Minnesota. Look for "Five-mile geodes" on the beaches and hills about five miles north of Grand Marais, Minnesota.

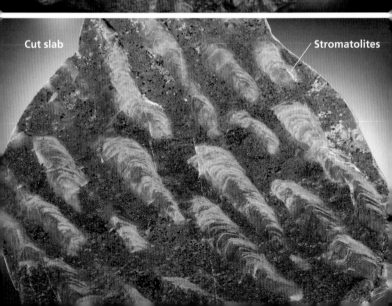

Rough specimen

Stromatolites

Cut slab

Stromatolites

Jasper, Mary Ellen

HARDNESS: Varies **STREAK:** N/A

Lake Superior Area
Occurrence

AREA: Northern Minnesota

ENVIRONMENT: Gravel pits and mines

WHAT TO LOOK FOR: Heavy, massive rock with distinct, banded column-like shapes within

SIZE: Mary Ellen jasper can be found in most any size.

COLOR: Browns, reds, yellows, greens, and blacks are all very common colors of Mary Ellen jasper.

OCCURRENCE: Rare

NOTES: Mary Ellen jasper is a very unique Minnesota collectible. These specimens of massive jasper contain two-billion-year-old fossils of ancient bacteria called stromatolites. Stromatolites formed mushroom-like or column-like colonies that rose upward out of bodies of water. Just like trees, each time these colonies grew, they formed a new layer. Mary Ellen jasper contains these ancient structures and shows the structure of these bacteria colonies.

Because Mary Ellen jasper is comprised mainly of chalcedony and quartz, it polishes extremely well and can be very valuable as display specimens. It occurs in a wide variety of colors, such as red, brown, green, or yellow, but the brighter and more complete the stromatolite specimens within the rock, the better. There is an extremely rare variety known as golden Mary Ellen jasper where the stromatolites within are a bright yellow, golden color.

WHERE TO LOOK: Mary Ellen Jasper can be found on Minnesota's Iron Range around Babbitt, Hibbing, and Virginia, Minnesota, and farther west in the Iron Range.

Lintonite

HARDNESS: 5-5.5 **STREAK:** Colorless

AREA: Minnesota's North Shore

Lake Superior Area
Occurrence

ENVIRONMENT: Lakeshore

WHAT TO LOOK FOR: Pale, glassy, irregular nodules (round rock clusters), greatly resembling beach-worn quartz

SIZE: Lintonite is primarily found as small pea-sized nodules (round rock clusters).

COLOR: Lintonite is a very pale green or blue.

OCCURRENCE: Common

NOTES: Lintonite is a common collectible on Minnesota's North Shore and can be very easy to find. As such, it is not very valuable, even in very fine specimens. Lintonite forms in other rocks, namely basalt, as small nodules (round rock clusters) within the gas bubbles in the rock. This mineral is often found associated with zeolites, namely the rare and very valuable thomsonite. Both minerals can sometimes be found in the same nodules (round rock clusters) or beach-worn nuggets, resulting in thomsonite appearing embedded within a piece of lintonite.

Some specimens of lintonite can be confused with small pieces of beach glass, but lintonite will appear more irregular, round, and less flat. In fact, due to its size and rounded, smooth shapes, lintonite often looks like jelly beans. The best places to find lintonite are any of Lake Superior's beaches north of Two Harbors. But don't fall prey to the popular myths that lintonite is "Minnesota jade."

WHERE TO LOOK: The best place to find Lintonite is around Cutface Creek, about four miles southwest of Grand Marais, Minnesota.

Rough specimen

Polished slab

Silkstone

HARDNESS: Varies   **STREAK:** N/A

Lake Superior Area
Occurrence

AREA: Minnesota's Iron Range

ENVIRONMENT: Gravel pits and mines

WHAT TO LOOK FOR: Fibrous, chatoyant (reflective) stones, often golden in color

SIZE: Varies from small stones to large boulders

COLOR: Silkstone's color can vary greatly, but the best specimens are golden yellow in color. Silkstone can also be brown or orange.

OCCURRENCE: Rare

NOTES: Silkstone, like binghamite, is a result of quartz replacing iron ores, namely hematite and goethite. Its colors can be bright but are generally not quite as intense as binghamite. Silkstone is most commonly yellow, gold, red, and brown but can also have shades of green, blue, and pink mixed throughout. And like binghamite, it can contain spots of quartz crystals and silver-black iron ore.

Binghamite and silkstone are so similar that they were, and sometimes still are, commonly referred to as being the same material. A geological study established the difference between the rocks. While binghamite has long, banded fibers and rich, intense colors, silkstone has short, wavy, and less organized mineral fibers with dullish colors.

Silkstone is highly sought after and valuable. Because it is quartz-based, it polishes very well and makes beautiful display specimens.

WHERE TO LOOK: Silkstone can be found on Minnesota's Iron Range, especially near Crosby, Minnesota.

Rough specimen

Taconite pellets

Taconite

HARDNESS: Varies **STREAK:** N/A

AREA: Minnesota

Lake Superior Area
Occurrence

ENVIRONMENT: Gravel pits and mine dumps

WHAT TO LOOK FOR: Dark, dense, and metallic rock

SIZE: Taconite can occur in gigantic masses, so loose pieces can run the full range of sizes.

COLOR: Dark gray to black with a slight metallic sheen

OCCURRENCE: Common

NOTES: If you try to look up taconite in any other rock book, odds are that you won't find it. The reason is that the word is a local term from Minnesota's Mesabi Iron Range and is simply used to describe any low-grade iron ore found in the area. Taconite, or iron ore, is generally a dark, slightly metallic metal but can contain other materials like jasper.

In the field, some forms of taconite may be hard to identify. But along railroad tracks or near the North Shore's industrial centers, taconite pellets are extremely common and easy to find. Taconite pellets are small balls of iron ore that are made by crushing raw taconite and using powerful magnets to extract the iron from the surrounding rock. The resulting iron is then melted and formed into easily transported taconite pellets that are about 65 percent iron. These pellets aren't valuable, nor are they very collectible, but they are easily identifiable and can be a souvenir of Northern Minnesota's primary industry.

WHERE TO LOOK: Taconite can be found in Minnesota's Iron Range around Babbitt, Hibbing, and Virginia, Minnesota, and farther west in the Iron Range. Taconite pellets can be easily found along railroads in the Two Harbors, Minnesota area.

Thomsonite in basalt

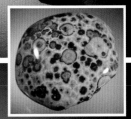

Small, polished specimens

Thomsonite

HARDNESS: 5-5.5 **STREAK:** Colorless

AREA: Minnesota's North Shore

ENVIRONMENT: Lakeshore

WHAT TO LOOK FOR: Pale pink, fibrous "eyes" within nodules (round rock clusters) in basalt

SIZE: Thomsonite is rarely found larger than penny-size, but pieces can be as large as walnuts.

COLOR: Pale pink to white is most common, but the best specimens have deep pinks and greens.

OCCURRENCE: Very rare

NOTES: Thomsonite is the most valuable and sought after of Lake Superior's many zeolites, the mineral family to which thomsonite belongs. One of the best places, if not the only place, to find thomsonite is the aptly named Thomsonite Beach of Northern Minnesota. Here, thomsonite can be found still embedded in basalt as amygdules (thomsonite which forms in round, gas bubble cavities). These specimens must be carefully and patiently removed from the stone using diamond carving tools to prevent damage to the brittle specimens.

One of Lake Superior's other zeolites, mesolite, is nearly identical to thomsonite and, in fact, cannot be distinguished from thomsonite by any means other than full laboratory testing. In that sense, some collectors' prized thomsonite may be mesolite.

Deep pinks and dark green "eyes" are the most valuable of thomsonite specimens.

WHERE TO LOOK: Look near Cutface Creek and Thomsonite Beach, both located four miles southwest of Grand Marais, Minnesota.

Wisconsin

Adventurers may view Lake Superior's geology near Wisconsin's Apostle Islands.

Despite having the least Lake Superior shoreline, Wisconsin is a great place to find collectible rocks and minerals.

With agates found on the shores near Superior and iron minerals found toward the Michigan border, not to mention the very rare concretions that can be found on the lake's shores near Ashland, Wisconsin has a lot to offer to rock hounds. And for a change of pace, you can take a ferry from Bayfield out to the Apostle Islands for a closer look at Lake Superior's unique geology.

Chalcopyrite

HARDNESS: 3.5-4 **STREAK:** Greenish-Black

Lake Superior Area
Occurrence

AREA: Primarily Wisconsin

ENVIRONMENT: Gravel pits and mines

WHAT TO LOOK FOR: Metallic, golden mineral with uneven faces

SIZE: Generally smaller specimens, no larger than a softball

COLOR: Chalcopyrite is a metallic yellow mineral resembling pyrite in structure but is more closely colored to gold than pyrite.

OCCURRENCE: Common

NOTES: Chalcopyrite is very often confused with pyrite because of its similar structure and color, but there are a few differences to look for. Chalcopyrite is a much richer yellow-gold color than pyrite, and its streak is more green than pyrite's. Although chalcopyrite's gold color is the most desirable, natural chalcopyrite is often tarnished black. Sometimes, this coating of tarnish can appear iridescent.

Chalcopyrite is a common basic copper ore associated with other copper minerals as well as copper itself. It can be found most easily where earth has been turned up, such as near a mine or in a gravel pit. It is also found within cavities in volcanic rock, such as in basalt and rhyolite.

WHERE TO LOOK: Chalcopyrite can be found in iron-mining areas like Hurley, Wisconsin.

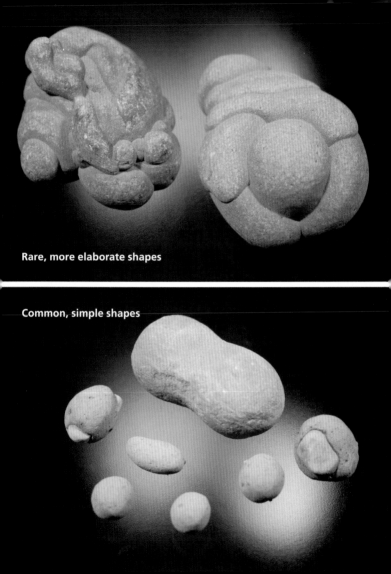

Rare, more elaborate shapes

Common, simple shapes

Concretions

HARDNESS: Varies **STREAK:** N/A

Lake Superior Area
Occurrence

AREA: Wisconsin's northern shores

ENVIRONMENT: Riverbanks and lakeshores

WHAT TO LOOK FOR: Hard, round, "sculpted" clay shapes

SIZE: Concretions range anywhere in size from grape-size up to softball-size.

COLOR: Concretions are always the color of the clay they formed in, primarily reds, browns, and tans.

OCCURRENCE: Rare

NOTES: In many areas, Lake Superior's banks are made almost entirely of clay, which is where concretions form. The eroding clay banks are ideal for the formation of concretions, which are hard, compact "sculptures" of solidified clay. These odd formations are highly collectible because of their strange and often exotic shapes.

Each concretion has a nucleus, such as a small stone or organic material that the clay forms around before hardening. This process is similar to how a pearl forms in an oyster. As the clay hardens around the object, it builds multiple layers upon it. Concretions are often found in the lake or on the clay banks after hard rains have washed away the top layers of clay.

Collecting concretions can be difficult as much of the land popular for concretion collecting is now privately-owned or is on Native American reservation land.

WHERE TO LOOK: Look for concretions on the shoreline east of Chequamegon Bay, Wisconsin, toward the Michigan border.

Michigan

Agates and datolites are gems found on Upper Michigan's vast beaches.

With its vast sandy beaches, rich history in mining, and immense geological diversity, Northern Michigan is a rock picker's heaven.

Lake Superior's southern shores hold some of the most valuable rocks and minerals in the region, many of which are a result of the huge copper deposits of the Keweenaw Peninsula. Whether you're hunting for agates near Grand Marais or precious metals in the Houghton area, Northern Michigan is one of Lake Superior's best places to hunt.

Rough specimens

Polished specimens

Agate, Brockway Mountain

HARDNESS: 7 **STREAK:** White

Lake Superior Area
Occurrence

AREA: Michigan

ENVIRONMENT: Lakeshore, gravel pits, mine dumps, and riverbanks.

WHAT TO LOOK FOR: Translucent, banded stones of varying colors, often with quartz crystals.

SIZE: Lake Superior agates can be any size, from tiny grains of sand up to softball-sized specimens

COLOR: Agates can run the full range of colors, but the most common colors are red, brown, gray, or white with clear, translucent quartz

OCCURRENCE: Rare

NOTES: At the northern end of Michigan's Keweenaw Peninsula is Mount Brockway, where you can find another unique variety of agate. These small agates are generally found as nodules (round rock clusters) no larger than two inches in diameter. Both their exterior and interior are colored in shades of orange, pink, gray, or pale blue, and the banding is most often opaque, greatly resembling paint agates.

WHERE TO LOOK: Brockway Mountain is about three miles southwest of Copper Harbor, Michigan, but many Brockway Mountain agates are found along Brockway Mountain Drive, which runs from Eagle Harbor, Michigan, to Copper Harbor, Michigan.

Agate, Copper Replacement

HARDNESS: 7 **STREAK:** White

AREA: Michigan

Lake Superior Area
Occurrence

ENVIRONMENT: Lakeshore, gravel pits, mine dumps, and riverbanks.

WHAT TO LOOK FOR: Translucent, banded stones of varying colors, often with quartz crystals.

SIZE: Lake Superior agates can be any size, from tiny grains of sand up to softball-sized specimens

COLOR: Agates can run the full range of colors, but the most common colors are red, brown, gray, or white with clear, translucent quartz

OCCURRENCE: Extremely Rare

NOTES: Of all the extremely rare, highly sought after, and very valuable agates in the Lake Superior region, none are outdone by the exquisite copper replacement agates found in Michigan's Keweenaw Peninsula. These specimens have been found no larger than in coin-sized nodules (round rock clusters), but even at such a small size, they've been known to fetch high prices.

Because they are a relatively new find, not much is known about how they form. It is hypothesized that after the agate nodules (round agate clusters) formed, a heavy copper solution seeped into the stone and replaced its delicate agate banding with copper. This is why the copper seems to mimic the banding of agates.

WHERE TO LOOK: These extremely rare agates can be found in copper mine dumps in the central part of Michigan's Keweenaw Peninsula, north of Houghton, Michigan.

Specimens courtesy of Karen Brzys

Specimens courtesy of Karen Brzys

Agate, Grand Marais

HARDNESS: 7 **STREAK:** White

Lake Superior Area
Occurrence

AREA: Michigan

ENVIRONMENT: Lakeshore, gravel pits, mine dumps, and riverbanks.

WHAT TO LOOK FOR: Translucent, banded stones of varying colors, often with quartz crystals.

SIZE: Lake Superior agates can be any size, from tiny grains of sand up to softball-sized specimens

COLOR: Agates can run the full range of colors, but the most common colors are red, brown, gray, or white with clear, translucent quartz

OCCURRENCE: Rare

NOTES: Not to be confused with Grand Marais, Minnesota, these agates from the Grand Marais, Michigan, area have been known to exhibit some unique traits not found elsewhere on Lake Superior. This area has vast expanses of sandy beaches, all prime for agate hunting, and you can spend hours upon hours here without seeing another person. A number of unique agate types have been found, including agates with black bands, copper inclusions (agates with copper intermixed within them) and agates with extravagant lace-like banding. A great place to see these agates is at the Gitchee Gumee Agate and History Museum in Grand Marais, Michigan, where many of these agates are on display.

WHERE TO LOOK: Look on the sandy beaches from Grand Marais, Michigan to Whitefish Point, Michigan.

Specimen courtesy of A.E. Seaman Mineral Museum

Specimen courtesy of A.E. Seaman Mineral Museum

Analcime

HARDNESS: 5-5.5 **STREAK:** White

AREA: Upper Michigan

Lake Superior Area
Occurrence

ENVIRONMENT: Gravel pits and mine dumps

WHAT TO LOOK FOR: Small, round, light-colored crystals on or in other rocks and minerals

SIZE: Crystals of analcime are generally very small and range from pea-size to walnut-size.

COLOR: Pure analcime is clear and colorless or white but can be tinted other colors, such as yellow, green, or red.

OCCURRENCE: Uncommon

NOTES: Analcime is one of the Lake Superior area's many types of zeolites. As a zeolite, it is commonly found within the cavities of another rock, specifically basalt or rhyolite. However, unlike many zeolites in the area, analcime sometimes occurs in complete crystals, and these can be found as small, faceted balls growing on the surface of other rocks and minerals. It is generally clear or white, but impurities can stain it other colors.

These glassy crystals can often be found with other collectible minerals like calcite, prehnite, and other zeolites. It is possible that analcime could be confused with quartz, though it is easy to tell them apart. Analcime is much softer than quartz and has a much different crystal structure. Quartz rarely, if ever, forms into small crystal balls like analcime.

WHERE TO LOOK: Mine dumps in Michigan's Keweenaw Peninsula are good places to look, as is the exposed rock around the L'Anse, Michigan area.

Small botryoidal
(grape-like) formations

Massive chrysocolla

Chrysocolla

HARDNESS: 2-4 **STREAK:** White

Lake Superior Area
Occurrence

AREA: Upper Michigan's Keweenaw Peninsula

ENVIRONMENT: Gravel pits and mine dumps

WHAT TO LOOK FOR: Blue to blue green masses in the matrix (the material other minerals form in) or botryoidal (grape-like) surfaces on a host rock

SIZE: Can form very large masses but is generally found in smaller pockets within other rock

COLOR: Chrysocolla is often turqouise-blue but can also be green.

OCCURRENCE: Uncommon

NOTES: The unique, bright green blue color of chrysocolla makes it a desirable gemstone and one commonly used in jewelry. Prized by collectors, Michigan's copper mines have produced stunning specimens of the mineral. This copper ore is most often found in mine dumps on Michigan's Keweenaw Peninsula as blue masses within other rock. Less frequently, it can be found as botryoidal (grape-like) structures within cavities of basalt or rhyolite.

As with many green or blue minerals in Upper Michigan, chrysocolla can be used as an indicator to find nearby copper deposits and often occurs with malachite, or oxidized copper (copper exposed to oxygen). Since it often combines with chalcedony, which makes it harder, chrysocolla can polish well and is sought after by collectors.

WHERE TO LOOK: Chrysocolla can be found in most of the copper mine dumps in the central area of Michigan's Keweenaw Peninsula, north of Houghton, Michigan.

Rough specimens

Specimen courtesy of A.E. Seaman Mineral Museum

Cut specimens

Specimen courtesy of A.E. Seaman Mineral Museum

Datolite

HARDNESS: 5-5.5 **STREAK:** Colorless

Lake Superior Area
Occurrence

AREA: Upper Michigan's Keweenaw Peninsula

ENVIRONMENT: Lakeshore, gravel pits, and mine dumps

WHAT TO LOOK FOR: Nodules (round rock clusters) with a cauli-flower-like outer texture with white to dark gray husk. Can be smooth and water-washed if found on the lakeshore.

SIZE: Size varies greatly, from pea-sized up to softball size but rarely larger.

COLOR: Most commonly, the interior is white or slightly pink in color, but it can be orange, gold, red, and rarely, blue and green.

OCCURRENCE: Uncommon

NOTES: Once just a pretty stone, datolites have now become a very valuable and sought after collectible. Though they can be found elsewhere in the United States, Lake Superior's datolites are primarily found in Michigan's Keweenaw Peninsula. While you can pick them on the lakeshore, most datolites are found in the old overburden piles left over from the old copper mines. As nodules, datolites form as irregular-shaped, cauliflower-like masses, generally an inch or two in diameter. The porcelain-like inside of a datolite can be many different colors, often white or pink. Rare and valuable stones are green and blue. Datolites are often found broken, making whole nodules (round rock clusters) semiprecious gems.

WHERE TO LOOK: Datolite can be found on the shores of Michigan's Keweenaw Peninsula, as well as in its mine dumps, but one of the best places to look is the very tip of the Keweenaw Peninsula, where there are few roads and navigation can be difficult.

Epidote crystals

Vesicular epidote

Beach-worn specimen

Epidote

HARDNESS: 6-7 **STREAK:** Gray

Lake Superior Area
Occurrence

AREA: Michigan's Upper Peninsula's North Shore

ENVIRONMENT: Lakeshore, gravel pits, and mine dumps

WHAT TO LOOK FOR: Generally, dark yellow green bladed crystals lining the insides of vesicles and other cavities in rock

SIZE: Epidote is generally found in small quantities, filling vesicles (cavities formed by gas bubbles) with small, long crystals.

COLOR: Epidote is most commonly dark green but can often be a very yellowish green.

OCCURRENCE: Common

NOTES: In its most true form, epidote is found in long, striated (grooved) yellow-green crystals, but in Michigan, most is found filling vesicles (cavities) in basalt with tiny, flat green crystals. It often occurs with calcite and prehnite, many times growing on or in them.

When epidote fills a vesicle (a cavity) completely or is found as a massive piece, it can look very much like chlorite, so much so that identifying it visually can be very difficult. While color can help—epidote often has more of a yellow hue while chlorite is a darker, almost blackish-green color—the primary difference between them is hardness. Epidote can't be scratched by a copper coin but chlorite can. And good specimens of epidote are far more valuable than even the best specimens of chlorite.

WHERE TO LOOK: The best crystal specimens are found in cavities of volcanic trap rock, such as basalt. It is easily obtained in the copper mine dumps of the central area of Michigan's Keweenaw Peninsula, north of Houghton, Michigan, and occurs in beach-worn specimens throughout the Upper Peninsula.

Hexagonal facets

Garnet

HARDNESS: 6.5-7.5 **STREAK:** Colorless

Lake Superior Area
Occurrence

AREA: Upper Michigan

ENVIRONMENT: Riverbeds and gravel pits

WHAT TO LOOK FOR: Dark, extremely heavy and angular crystals with definite multisided structures

SIZE: Can range widely in size but is generally pea-sized up to walnut-sized

COLOR: Garnets found in Michigan are often black, gray, or reddish brown.

OCCURRENCE: Rare

NOTES: Garnet is a metamorphic mineral, meaning it only forms under great heat and pressure. Naturally, there are many kinds of garnet due to the rocks and minerals in which it forms. Garnets run the full range of colors worldwide, but garnets from Michigan are dark crystals, oftentimes red or brown but also commonly black or gray. Good examples of garnet have a hexagonal crystal structure and exhibit very definite crystal planes; however, poor specimens can be far less distinct.

Garnets can most easily be found in places where fresh earth is being moved, such as riverbeds. They are extremely dense and feel very heavy for their size, making them easy to identify. These garnets are relatively rare, and exceptional specimens are both highly collectible and valuable.

WHERE TO LOOK: These rare crystals can be found around Marquette and Ishpeming, Michigan.

Short grunerite crystals

Specimen courtesy of A.E. Seaman Mineral Museum

Specimen courtesy of A.E. Seaman Mineral Museum

Grunerite

HARDNESS: 5-6 **STREAK:** Colorless

Lake Superior Area
Occurrence

AREA: Michigan

ENVIRONMENT: Gravel pits and mine dumps

WHAT TO LOOK FOR: Small, prismatic crystals with a pearly luster, often occurring within another rock

SIZE: Grunerite crystals are mostly small, being no larger than a dime.

COLOR: Crystals of grunerite are mostly brown, green, or gray.

OCCURRENCE: Common

NOTES: Grunerite is a small variety of crystal often found embedded within another rock. Its crystals are short, fibrous, and columnar, generally formed in large groups, all pointing the same direction within its parent rock. Golden brown grunerite crystals can often appear to be organic and almost seed-like.

Grunerite is not particularly collectible or valuable except to those looking to expand their collection. Grunerite can also be purchased commercially as a decorative stone called amosite.

WHERE TO LOOK: Grunerite can be found in a number of locations, but is primarily found in inland regions of the Upper Peninsula where there is exposed rock.

Copper

Silver

Silver

Halfbreed

HARDNESS: 2.5-3 **STREAK:** N/A

Lake Superior Area
Occurrence

AREA: Michigan's Keweenaw Peninsula

ENVIRONMENT: Primarily mine dumps

WHAT TO LOOK FOR: Copper and silver in the same specimen

SIZE: Can be a wide range of sizes, though one of the metals in the specimen is generally smaller than the other

COLOR: Halfbreed exhibits the color of both copper and silver. When oxidized (changed due to exposure to oxygen), these specimens can be dark green or black.

OCCURRENCE: Rare

NOTES: Halfbreeds are a unique combination of copper and silver in the same specimen. In halfbreeds, these two valuable metals have formed onto each other and can be a very interesting addition to your collection. Many times, one of the metals in the specimen is more prevalent than the other. Most of the time, it is copper with a small amount of silver grown onto it, though there have been examples where copper formed on a larger piece of silver.

Halfbreeds are usually found in the old mine dumps in the Keweenaw Peninsula of Michigan, where most other copper and silver specimens can be found. When hunting, a metal detector is the best tool for finding them. And if you find a piece of copper, look it over carefully for a piece of silver, oxidized to a dark gray (changed due to exposure to oxygen).

WHERE TO LOOK: The copper mine dumps of the central area of Michigan's Keweenaw Peninsula, north of Houghton, Michigan, are the best places to look.

Limestone

Coral structure

Horn Coral

HARDNESS: Varies **STREAK:** N/A

Lake Superior Area
Occurrence

AREA: Upper Michigan

ENVIRONMENT: Lakeshore

WHAT TO LOOK FOR: Light-colored stones that contain segments of distinctly organic structures

SIZE: Most pieces of horn coral are no larger than your thumb.

COLOR: Horn coral specimens can be white, beige, or yellow.

OCCURRENCE: Uncommon

NOTES: Michigan's northern shores hold some of the most easily found fossils in the Lake Superior region. These fossils are known as horn coral and can be very interesting pieces to add to your collection, especially as fossils are not very common in the area. These stones consist mostly of lightweight white or tan limestone with odd-looking inclusions (minerals that formed within it) that can sometimes resemble wood grain or the veins on a leaf. These strange structures are horn coral fossils.

Lake Superior was once the site of an ancient sea, and much later on, a vast marshland. Vast beds of limestone, known for preserving fossils, formed in this area. When the glaciers of the many ice ages came through the region, they scraped up and pulverized the limestone, depositing it all around the region. Michigan is one of the only places to find these specimens. Most of the time, horn coral fossils are only really visible when wet, making the water's edge the best place to look.

WHERE TO LOOK: The sandy beaches surrounding Grand Marais, Michigan are great places to find specimens.

Rough specimens

Polished specimen

Kona Dolomite

HARDNESS: 4 **STREAK:** N/A

Lake Superior Area
Occurrence

AREA: Upper Michigan

ENVIRONMENT: Gravel pits and mines

WHAT TO LOOK FOR: Mottled, light-colored rock in shades of pinks, browns, and creams

SIZE: Kona dolomite occurs massively and can be very large as well as occur as small stones.

COLOR: Shades of pink, brown, red, yellow, and cream are the most common colors of Kona dolomite.

OCCURRENCE: Uncommon

NOTES: The name Kona dolomite refers to a variety of massive dolomite, or dolostone, that occurs in the Kona Hills region of Upper Michigan. This rock can be a wide range of colors, but mostly, it occurs in shades of brown or pink and also can be cream-colored or white. It has a cracked appearance, which contributes to its mottled colors and unique look. Kona dolomite isn't particularly valuable, but nicely colored specimens can be quite collectible. The stone has been used for years as a carving and jewelry material.

Kona dolomite contains traces of fossilized remains of ancient bacteria, called stromatolites. The bacteria has been dated at more than two billion years old. Kona dolomite can be found around Marquette County, Michigan, where it is quarried.

WHERE TO LOOK: The Kona Hills area is about eight miles southwest of Marquette, Michigan, in the central portion of Michigan's Upper Peninsula. The Lindberg Quarry near Marquette, Michigan has excellent material.

Limestone

HARDNESS: 3-4 **STREAK:** N/A

Lake Superior Area
Occurrence

AREA: Primarily Michigan

ENVIRONMENT: Lakeshore

WHAT TO LOOK FOR: Light-colored, soft stones that feel slightly chalky to the touch

SIZE: Limestone occurs massively, but on the beaches, it is mostly found as small, water-worn pebbles.

COLOR: White or cream colors are the most common, but many specimens contain dark gray inclusions (minerals that formed within it).

OCCURRENCE: Very common

NOTES: Before Lake Superior was a huge body of water, it was a vast marshland, which led to the formation of an extremely large bed of limestone. Because limestone is very soft, the glaciers that moved through the area gouged out the limestone, leaving the lake behind. Most of the limestone was scraped away and brought far south of the lake, but many beach-worn remains can be found on the northern shores of Michigan's Upper Peninsula. This soft, chalky white stone can be easily found and identified. And if there is ever a question whether or not a specimen is limestone or not, a simple test of pouring vinegar on the sample will yield an answer. If it is limestone, the vinegar will cause the stone to fizz and bubble. Limestone often contains organic material, including fossils that once inhabited this area millions of years ago.

WHERE TO LOOK: Upper Michigan's northern shores are all good places to look, but the sandy beaches from Grand Marais, Michigan to Whitefish Point, Michigan are great places to look.

Specimen courtesy of A.E. Seaman Mineral Museum

Manganite

HARDNESS: 3.5-4 **STREAK:** Brown or Black

Lake Superior Area
Occurrence

AREA: Upper Michigan

ENVIRONMENT: Mine dumps

WHAT TO LOOK FOR: Black, metallic, fibrous splinters radiating out from a central point

SIZE: Manganite crystals are generally short, no longer than an inch or two long.

COLOR: Manganite is metallic black and has a silver luster.

OCCURRENCE: Uncommon

NOTES: Manganite is a form of oxidized manganese (manganese exposed to oxygen) often associated with mines, gravel pits, and wherever earth has been moved. These short, splinter-like crystals grow in a radiating structure (spread out from a central point), forming fan shapes. It is always a steel gray or black color and appears metallic and reflective. The crystals are very brittle, and complete unbroken specimens are rare and very collectible.

Manganite greatly resembles pyrolusite in appearance, but as with many other look-alike minerals, there are simple tests to differentiate between the two. Manganite is not as hard as pyrolusite, but an average scratch test may be difficult, since both minerals are fibrous and brittle in nature. Instead, a streak test may be easier. Manganite has a reddish brown to black streak, whereas pyrolusite has a bluish black streak. Even the streaks, however, can both occasionally appear black.

WHERE TO LOOK: Manganite is associated with iron and copper mining and can be found in mine dumps all over the Keweenaw Peninsula and surrounding areas.

Bladed marcasite crystals

Specimen courtesy of A.E. Seaman Mineral Museum

Marcasite

HARDNESS: 6-6.5 **STREAK:** Dark Green to Brown

AREA: Upper Michigan

Lake Superior Area
Occurrence

ENVIRONMENT: Gravel pits and mine dumps

WHAT TO LOOK FOR: Small, metallic, bladed crystals growing on the surface of other rocks and minerals

SIZE: Marcasite crystals are often small, pea-sized specimens but can be larger.

COLOR: Marcasite has a color similar to pyrite in that it is metallic and golden, but unlike pyrite, marcasite tends to have a greenish tint.

OCCURRENCE: Uncommon

NOTES: Marcasite is a golden metallic crystal similar to chalcopyrite or pyrite. One of the primary differences is its greenish hue and thin, bladed crystals. These crystals are often referred to as having a "cockscomb" shape, meaning it resembles the top of a rooster's head. As such, these crystals appear thin with a jagged top edge.

Marcasite is heavy and fragile and is often found growing on the surface of other minerals as small crystals. However, this is the ideal appearance for marcasite. More often, it is heavily oxidized (exposed to oxygen) with patches of white dust on the surface. This oxidization (a chemical change due to exposure to oxygen) occurs relatively quickly and causes the mineral to crumble. Marcasite is an iron-based mineral and is found alongside iron ores. It is very collectible and fine specimens can be valuable.

WHERE TO LOOK: Look for marcasite in the iron mine dumps of northern Michigan's Iron Range.

Mohawkite

Specimens courtesy of A.E. Seaman Mineral Museum

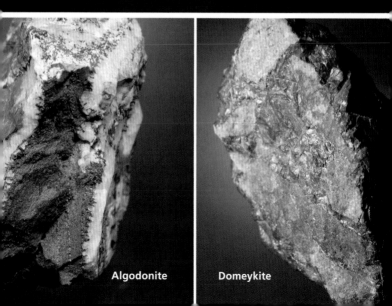

Algodonite

Domeykite

Mohawkite

HARDNESS: Varies **STREAK:** N/A

AREA: The northern side of Michigan's Keweenaw Peninsula

ENVIRONMENT: Mine dumps

WHAT TO LOOK FOR: A heavy, metallic stone resembling copper or bronze, often within other rocks and minerals

SIZE: Mohawkite generally occurs in veins or as an inclusion (a mineral which formed in another rock), therefore making most specimens no larger than softball-sized.

COLOR: In its purest form, Mowakhite resembles bronze or copper.

OCCURRENCE: Rare

NOTES: Mohawkite is a very rare, unique mineral found only on the Keweenaw Peninsula. It was originally found in the dumps of the copper mines around the towns of Mohawk and Ahmeek, for which it was named. Mohawkite consists of two minerals, algodonite and domeykite, both of which are copper arsenides, meaning they contain arsenic. Because of this fact, you may not want to handle these minerals very often.

All three minerals are very similar, especially in color. Domeykite and algodonite are a bronze-like color and thus mohawkite is as well. However, the most sought after specimens of mohawkite contain flecks of quartz. This variety is known as snowflake mohawkite and is very valuable in large specimens.

WHERE TO LOOK: Mohawkite and its related minerals are only found in mine dumps around Mohawk and Ahmeek, Michigan, about sixteen miles northeast of Houghton, Michigan, in the central area of the Keweenaw Peninsula.

Rough specimens

High-grade polished specimens

Pumpellyite (Greenstone)

HARDNESS: 6 **STREAK:** White

AREA: Michigan's Keweenaw Peninsula and Isle Royale

ENVIRONMENT: Lakeshore and mine dumps

WHAT TO LOOK FOR: Dark, irregular nodules (round rock clusters) in trap rock, very similar in appearance to chlorite nodules (round chlorite clusters)

SIZE: Pumpellyite generally remains very small, no bigger than a dime.

COLOR: Dark to light green nodules (round rock clusters) that have a distinct "turtle-back" pattern once broken or polished.

OCCURRENCE: Very rare

NOTES: Most people know pumpellyite by its more common name, greenstone, or more specifically, Isle Royale greenstone. It forms as nodules (round rock clusters) in the gas bubbles within volcanic trap rock (rock which traps minerals in its gas bubbles), primarily basalt. The best specimens come from Isle Royale, where it can be found as beach pebbles, but it can also be found in the mine dumps on the northern tip of the Keweenaw Peninsula.

Pumpellyite is a highly prized and very valuable gemstone, widely used in jewelry and as specimens. Its turtle-back pattern can fetch high prices for very fine examples. While the best specimens are found on Isle Royale, it must be noted that collecting the stone from Isle Royale is illegal. The reason is that Isle Royale is now a national park, and you cannot legally remove anything from the island or surrounding water.

WHERE TO LOOK: Pumpellyite can be found in the copper mine dumps of the upper portion of the Keweenaw Peninsula. Isle Royale is also a very good place to see pumpellyite, but it is illegal to collect from the island.

Specimen courtesy of A.E. Seaman Mineral Museum

Asbestos

Specimen courtesy of A.E. Seaman Mineral Museum

Serpentine

HARDNESS: 3-5 **STREAK:** White

Lake Superior Area
Occurrence

AREA: Upper Michigan

ENVIRONMENT: Mine dumps and gravel pits

WHAT TO LOOK FOR: Light-colored fibrous crystals with a very waxy feel

SIZE: Serpentine can occur in large masses, but generally, specimens are the size of your hand or smaller.

COLOR: Michigan serpentine generally occurs in shades of greens and yellows.

OCCURRENCE: Uncommon

NOTES: There are many kinds of serpentine all over the world, but the type found in Michigan is known as crysotile serpentine. This waxy, striated (grooved) crystal is a very collectible mineral that can be a unique addition to any collection. This mineral can become very fibrous and be loosely held together. At this stage, it greatly resembles fabric or fur and becomes very fragile. This "fuzz" is more commonly known as asbestos and can be very dangerous if inhaled. When handling crysotile serpentine, be very careful not to loosen and separate the fibers because they very easily become airborne.

This mineral has been widely mined and used for many industrial fireproofing applications. However, its use has been drastically reduced after the discovery of its hazardous nature. Take care with the specimens you've collected to avoid inhaling particles or dust.

WHERE TO LOOK: Serpentine can be found in some of the copper mine dumps of the Keweenaw Peninsula.

Silver

HARDNESS: 2.5-3 **STREAK:** Shiny White to Gray

Lake Superior Area
Occurrence

AREA: Northern Michigan's Keweenaw Peninsula

ENVIRONMENT: Mine dumps

WHAT TO LOOK FOR: Small, dark, sharp pieces of malleable metal

SIZE: Usually, it is very small, found as small flakes or veins within a host rock.

COLOR: Natural silver will most likely be oxidized (changed due to exposure to oxygen) black or gray but can be cleaned to exhibit the characteristic metallic silver color.

OCCURRENCE: Rare

NOTES: Silver is one of the Keweenaw's most highly collectible minerals. Like many other valuable minerals found in Upper Michigan, silver is best found in the mine dumps by using a metal detector. Many times, it is found as small flakes within other rock, but more rarely, it can be found as loose nuggets or crystals. Silver crystals are the most valuable type of silver due to their rarity. They occur when the silver formed within a cavity large enough for it to grow into its natural structure.

Most silver specimens are very small and delicate, making finding them very difficult. And since silver occurs mostly where copper does, using a metal detector can yield mixed results. Copper can be found in much larger pieces than silver, so a metal detector will most likely detect a piece of copper before it does silver.

WHERE TO LOOK: Silver can be found in the mine dumps and riverbeds of the Keweenaw Peninsula.

Embedded staurolite crystal

Staurolite cross

Staurolite

HARDNESS: 7-7.5 **STREAK:** White

AREA: Upper Michigan

Lake Superior Area
Occurrence

ENVIRONMENT: Gravel pits

WHAT TO LOOK FOR: Small, dark rectangular crystals often embedded in rock

SIZE: Staurolite crystals generally remain small, no bigger than your thumb.

COLOR: Crystals of staurolite range from black to reddish brown.

OCCURRENCE: Uncommon

NOTES: Staurolite, much like garnet, is formed metamorphically under great heat and pressure. As a result, it is often found firmly embedded within the schist it formed in. When removing them from their parent rock, take care to prevent damaging the staurolite crystals. Staurolite is also often found as loose crystal.

The best specimens of staurolite will have a distinctive cross shape. Staurolite crystals will often be twinned, with the second crystal extending through the center of the first, resembling a cross, or X-shape. This formation is the most collectible of all staurolites, and fine specimens can be valuable.

Like garnets and other metamorphic crystals, staurolites are dense, heavy and rarely very big. Most staurolites are no bigger than your thumbnail. They are also quite hard, as hard as and sometimes harder than quartz.

WHERE TO LOOK: Look for staurolite near the Peavy Dam on the Michigamee River in Michigan's Upper Peninsula.

Green surface oxidization
(a chemical change due to exposure to oxygen)

Specimen courtesy of A. E. Seaman Mineral

Tenorite

HARDNESS: Varies **STREAK:** N/A

AREA: Upper Michigan

Lake Superior Area
Occurrence

ENVIRONMENT: Mine dumps and gravel pits

WHAT TO LOOK FOR: Dark, dense stone with a dull, flat luster

SIZE: Tenorite can occur massively and therefore occurs in a wide range of sizes.

COLOR: The color of tenorite is primarily gray or black, but it can contain spots of blue or green oxidization (a chemical change due to exposure to oxygen).

OCCURRENCE: Uncommon

NOTES: Michigan's Upper Peninsula is so rich with copper that many rocks and minerals in the area are affected by or are a direct result of its presence in the earth. Tenorite is a perfect example of this process. It consists entirely of oxidized (exposed to oxygen) copper in a massive form. It is dark and dense and not much to look at but is often sought after by collectors interested in copper minerals. As a massive rock, it can be found in a wide range of sizes and primarily in mine dumps or other places where earth has been moved.

Tenorite is often associated with other copper-related minerals, such as malochite and chrysocolla. Many times, these other minerals are actually formed on pieces of tenorite. Finding tenorite can also be misleading due to the high copper content. While it is normally dull gray or black, it can have an oxidized layer of green or blue (a layer altered due to exposure to oxygen) on its surface.

WHERE TO LOOK: Tenorite is primarily found in the mine dumps of the Keweenaw Peninsula.

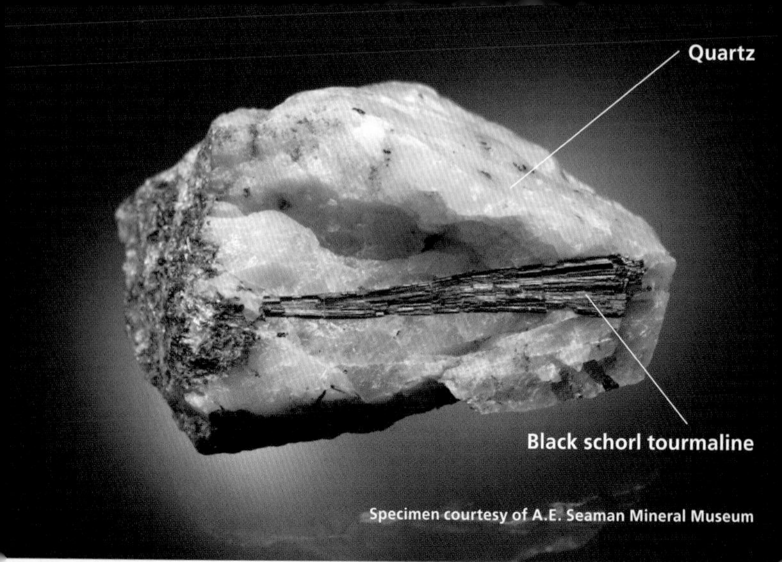

Quartz
Black schorl tourmaline

Specimen courtesy of A.E. Seaman Mineral Museum

Mica
Small tourmaline fragments

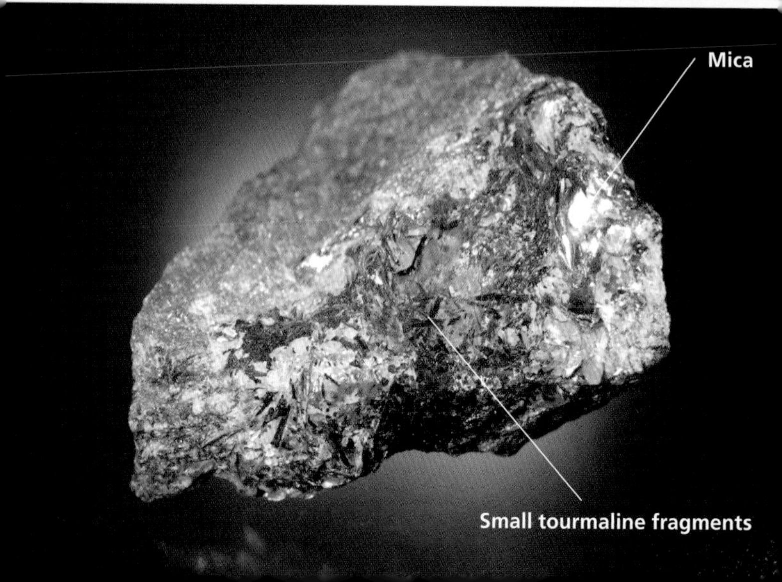

Tourmaline

HARDNESS: 7-7.5 **STREAK:** Colorless

Lake Superior Area
Occurrence

AREA: Upper Michigan

ENVIRONMENT: Gravel pits and mine dumps

WHAT TO LOOK FOR: Long, black, striated (grooved) crystals growing within other minerals

SIZE: Tourmaline is generally found in small crystals, but specimens can rarely be found as long as pencils.

COLOR: Tourmaline is most commonly a deep, rich black.

OCCURRENCE: Uncommon

NOTES: Tourmaline is a crystal found around the world and in many different colors. In Michigan, the black variety, called schorl, is associated with quartz and other minerals. Tourmaline is found as long, slender crystals with a very characteristic, striated (grooved), fibrous texture, and cross sections of well-formed specimens will reveal that the mineral is generally triangular. Tourmaline can be found where granite and other coarse-grained rocks are present, and many times within them as part of their mineralogical structure. Tourmaline is also commonly found growing in the interior or on the exterior of quartz crystals. Tourmaline is highly collectible and can be very valuable if a very complete, intact specimen is found. Schorl, or black tourmaline, is colored by the high content of sodium and iron within the mineral. This makes Michigan's tourmaline very hard, heavy, and brittle.

WHERE TO LOOK: Tourmaline can be found in places there is exposed rock, such as mine dumps and gravel pits and is often found within quartz or granite. Look near Marquette, Michigan.

Beach-worn specimens

Unakite

HARDNESS: 6-7 **STREAK:** N/A

Lake Superior Area
Occurrence

AREA: Michigan's Upper Peninsula's North Shore

ENVIRONMENT: Lakeshore

WHAT TO LOOK FOR: Small beach-worn nuggets made of two distinct minerals of varying colors, primarily green and pink

SIZE: Pieces of unakite range from pea-sized to softball-sized but are usually no bigger than a walnut.

COLOR: Unakite always occurs in shades of green mixed with reds or oranges.

OCCURRENCE: Common

NOTES: Unakite is a unique stone in that it consists of two colorful minerals swirled together into one collectible rock. The two minerals within unakite are green epidote and orange or pink feldspar, primarily orthoclase or microcline feldspar. This combination forms within bodies of granite and can be found all over the world. Michigan's unakite is most often found on the beaches of the Upper Peninsula as small, smooth nuggets. It is very common and extremely easy to find and identify once you learn its colors. Unakite is very collectible but not very valuable, and it has been used for years as a carving and jewelry material. Even the most tiresome day of beachcombing will at least yield some nice specimens of unakite, if nothing else, since it can be found all over Lake Superior's Michigan shorelines.

WHERE TO LOOK: Unakite is primarily found on the beaches of northern Michigan as rounded pebbles. One can find it all over the Upper Peninsula, from the Keweenaw Peninsula all the way to Whitefish Point, Michigan.

Glossary

AMYGDULE: Material formed in round, gas bubble cavities within igneous rocks

BOTRYOIDAL: Mineral formations that resemble a bunch of grapes

BRECCIA: Coarse, sharply broken rock that has been cemented back together by a finely grained mineral

CALCIC: Rich in calcium

CANADIAN SHIELD: An enormous bedrock formation stretching from Eastern to Central Canada and extending north from the Great Lakes up to the Arctic Ocean

CHALCEDONY: Microcrystalline quartz formations in which the quartz structures are invisible to the naked eye

CHATOYANT: Refers to the way light enters a mineral in which it appears to move and reflect differently

CONCENTRIC: A formation having a common center point, as in a target

CONCHOIDAL: Refers to a curved structure resembling a half-moon shape, a common fracture in quartz-based minerals

CONCRETION: A hard, round, compact mass of rock formed by precipitation of minerals around a nucleus

FAULTED AGATE: Agate that has gone through an event causing a crack to occur, resulting in an alignment shift in the banding

FIBROUS: Refers to minerals that appear thread-like

FOLIATED: Thin sheets of material arranged in layers, as in mica

GRANULAR: Rocks made up of mineral particles coarser than sand but finer than pebbles

IGNEOUS: Volcanic rock formed by the cooling of molten rock

INCLUSION: A mineral that formed within another

IRIDESCENT: The property of a mineral to change color through the refraction of light within the mineral

LAVA: Molten rock that has reached the Earth's surface

MAGMA: Molten rock that is still deep inside the Earth

MASS: A large concentration of a rock or mineral

MATRIX: The finely grained, massive material that other minerals form in

METAMORPHIC: A rock that has been changed due to great heat and pressure

MICA: A group of glass-like minerals that form foliated, or layered, crystal. Micas are a prominent rock-forming mineral

MICACEOUS: Containing or consisting of mica

MICROCRYSTALLINE: Crystals that cannot be seen with the unaided eye and require a microscope

NODULE: A small rock or mineral cluster, generally in a round shape

ONYX: A variety of banded agate or chalcedony with straight, parallel bands

OPAQUE: A mineral that light cannot pass through

PEGMATITE: Igneous rock with an extremely large grain size

PILLOW LAVA: Lava that cooled quickly under water and formed smooth pillow-like shapes

PLATY: Flat, thin, and plate-like

PORPHYRY: Larger crystals that formed within igneous rocks

PYRAMIDAL: Pyramid-like shape

RADIAL: Having a central point of origin

RIFT: A long, narrow trough created by the Earth's surface parting and splitting

ROCK: A solid that is made up of multiple minerals

SEDIMENTARY ROCK: Rock formed by the solidification of sediments

SEDIMENT: Solid material that has settled due to wind or water movement

SCHIST: The final stage of metamorphism with well-developed, parallel layers of minerals

SILICA: Silicon dioxide, a quartz-forming material

SLAG GLASS: Glass left over from smelting processes, formed by melting rock which cooled very rapidly.

SMELTING: Melting or firing an ore as to separate the rock from the metal

SPECIMEN: A sample, as in a piece of a mineral

TABULAR: A mineral with a thickness much less than that of its other dimensions

TRANSLUCENT: Partly transparent; a material capable of letting through some light

TRANSPARENT: Clear; something which light can pass through

TRAP ROCK: Vesicular igneous rock that traps other minerals in its gas bubbles

VESICLE: A cavity formed by a gas bubble in a volcanic rock

VESICULAR: A mineral with vesicles (cavities formed by gas bubbles)

VITREOUS: A mineral with the appearance of glass

Lake Superior Rock Shops and Museums

Minnesota

3M MUSEUM
201 Waterfront Drive
Two Harbors, MN 55616
(218) 834-4898

AGATE CITY ROCKS AND GIFTS *(the authors' store)*
721 7th Avenue (Highway 61)
Two Harbors, MN 55616
(218) 834-2304
www.agatecity.com

BEAVER BAY AGATE SHOP
1003 Main Street
Beaver Bay, MN 55601
(218) 226-4847

FRAGMENTS OF HISTORY
1 1/2 West Superior Street
Duluth, MN 55802
(218) 786-0707

MOOSE LAKE STATE PARK
Agate and Geological Interpretive Center
Located 1/2 mile east of I-35 at the Moose Lake exit
(218) 485-5420, (866) 857-2757

Wisconsin

JACK PINE ROCK SHOP
15822 East 2nd Street
Hayward, WI 54843
(715) 934-2130

Michigan

GITCHE GUMEE AGATE AND HISTORY MUSEUM
P.O. Box 308, East 21739 Brazil Street
Grand Marais, MI 49839
(906) 494-2590

KEWEENAW GEM AND GIFT, INC.
1007 West Memorial
Houghton, MI 49931
(906) 482-8447

NATURE'S PICKS ROCK SHOP
600 Cloverland Avenue
Ironwood, MI 49938
(906) 932-7340

PROSPECTOR'S PARADISE
P.O. Box 86
Mohawk, MI 49950
(906) 337-6889

ROCK KNOCKER'S ROCK SHOP
490 North Street (U.S. 41)
Ishpeming, MI 49849
(906) 485-5595

SWEDES
P.O. Box 66, U.S. 41
Copper Harbor, MI 49918
(906) 289-4596

A.E. SEAMAN MINERAL MUSEUM
Michigan Technological University
1400 Townsend Drive
Houghton, MI 44931-1295
(906) 487-2572
www.museum.mtu.edu

Bibliography and Recommended Reading

Books about the Lake Superior Region

Brzys, Karen, *Understanding and Finding Agates*. Hancock: Book Concern Printers, 2004.

Carlson, Michael. *The Beauty of Banded Agates*. Edina: Fortification Press, 2002.

Marshall, John. *The "Other" Lake Superior Agates*. Beaverton: Llao Rock Publications, 2003.

Pabian, Roger, et al. *Agates: Treasures of the Earth*. Buffalo: Firefly Books Limited, 2006.

Robinson, Susan. *Is This an Agate?* Hancock: Book Concern Printers, 2001.

Stensaas, Mark "Sparky," *Rock Picker's Guide to Lake Superior's North Shore*. Duluth: Kolath-Stensaas Publishing, 2000.

Wolter, Scott, *The Lake Superior Agate*. Edina: Burgess Publishing Company, 1986.

Zeitner, June Culp, *Midwest Gem, Fossil and Mineral Trails of the Great Lakes States*. Baldwin Park: Gem Guides Book Company, 1999.

General Reading

Chesteman, Charles W., *The Audubon Society Field Guide to North American Rocks and Minerals*. New York: Knopf, 1979.

Mottana, Annibale, et al. *Simon and Schuster's Guide to Rocks and Minerals*. New York: Simon and Schuster, 1978.

Pellant, Chris, *Rocks and Minerals*. New York: Dorling Kindersley Publishing, 2002.

Pough, Frederick H., *Rocks and Minerals* Boston: Houghton Mifflin, 1988.

Index

About the Authors

Bob Lynch has been working with rocks since 1973, when he wanted more variation in stones for his work with jewelry. When he moved to Two Harbors, Minnesota, in 1982, he was opened up to a new world of minerals. In 1992, Bob and wife Nancy acquired Agate City Rock Shop, a family business founded by Art Rafn in 1962. Bob can be found daily in his shop, now called Agate City Rocks and Gifts, on Highway 61 in Two Harbors.

Dan Lynch has a bachelor of fine arts degree in graphic design and photography but as the son of a rock hound he has always had an interest in rocks and minerals. With a love of writing and illustrating, Dan is pursuing a career in the publication industry. Dan currently lives in Duluth, Minnesota, with his fiancée Julie.

Notes

Notes